Pedestrian Accident Reconstruction

Pedestrian Accident Reconstruction

Nathan A. Rose
Principal Accident Reconstructionist
Explico Inc.
nathan@explico.com

Neal Carter
Principal Engineer
Explico Inc.
neal@explico.com

400 Commonwealth Drive
Warrendale, PA 15096-0001 USA
E-mail: CustomerService@sae.org
Phone: 877-606-7323 (inside USA and Canada)
           724-776-4970 (outside USA)
Fax: 724-776-0790

Library of Congress Catalog Number 2025911728
http://dx.doi.org/10.4271/9781468609714

ISBN-Print   978-1-4686-0970-7
ISBN-PDF   978-1-4686-0971-4
ISBN-epub 978-1-4686-0972-1

To purchase bulk quantities, please contact: SAE Customer Service

E-mail: CustomerService@sae.org
Phone: 877-606-7323 (inside USA and Canada)
           724-776-4970 (outside USA)
Fax: 724-776-0790

Visit the SAE International Bookstore at books.sae.org

Publisher
Sherry Dickinson Nigam

Product Manager
Amanda Zeidan

Production and
Manufacturing Associate
Michelle Silberman

Contents

Chapter 05 – Event Data Recorders in Pedestrian Accident Reconstruction

Chapter 06 – Human Factors and Nighttime Visibility

Preface

Why the need for a book on pedestrian accident reconstruction? Other books exist that describe methods for reconstructing pedestrian collisions. Those books are relevant and useful, but they were written prior to the ubiquitousness of electronic, video, and audio evidence. In addition, they were written when the use of simulation software was less common in accident reconstruction. For example, at the time of the second edition of *Pedestrian Accident Reconstruction and Litigation*, in 1999, Jerry Eubanks observed that "computer programs are becoming the 'in' thing with accident reconstructionists…Some expensive computer programs are available for use in mainframe computers to analyze pedestrian collisions."

This book describes accident reconstruction methods for a new era, where event data from vehicle sensors and video evidence have become centerpieces of accident reconstruction. Simulation is also used extensively in this book. While simulation is not always needed when reconstructing a pedestrian collision, its use can be invaluable. Many concepts and principles are readily illustrated with simulation. The same concepts and principles often cannot be illustrated with staged collisions or real-world pedestrian impacts, largely because the

systematic parameter variations that are easily accomplished in simulation are simply not possible with staged collisions or in a real-world collision setting. The primary simulation tool used in this book is the PC-Crash multibody pedestrian model. PC-Crash is a widely used simulation software package, but it is not the only one. PC-Crash is what we have often used in our daily practice, and so it is used in this book. But similar illustrations could have been created with other simulation tools, and our purpose here is not to advocate for one simulation software or another.

That said, the PC-Crash multibody pedestrian model is a robust model that has been validated. Chapter 3 details this validation, but the model is used throughout the book. The multibody model consists of a series of rigid bodies representing the various body parts (the head, torso, and pelvis, for example) connected with joints. The user can specify the weight and height of the pedestrian, the initial position and orientation each body part (within the constraints imposed by the joints), and the initial velocity of the pedestrian. The software will sense the contact between the pedestrian's body parts and the vehicle or the ground in the simulation and calculate the forces and resulting motion for the pedestrian. This model is particularly useful when other methods are not feasible. For example, empirical throw distance equations are generally only applicable to forward projections and wrap trajectories, not to fender vaults, roof vaults, or sideswipe pedestrian collisions.

So, our intention with this book is to document the advances and changes in the methods for pedestrian accident reconstruction that been developed and implemented since the publication of other popular books on the topic. This topic is extensive, and this book is far from exhaustive.

Nathan Rose
Parker, Colorado
March 2025

List of Acronyms

3DEP - Three-Dimensional Elevation Program

ABS - Anti-Lock Brake Systems

ACMs - Airbag Control Modules

ADTs - Anthropomorphic Test Devices

AEB - Automatic Emergency Braking

ASPRS - American Society for Photogrammetry and Remote Sensing

CDR - Crash Data Retrieval

CG - Center of Gravity

CISS - Crash Investigation Sampling System

DCT - Discrete Cosine Transformation

DVR - Digital Video Recorder

ECM - Engine Control Module

ECU - Electronic Control Unit

EDRs - Event Data Recorders

FCW - Forward Collision Warning

FFD - Freeze Frame Data

FPS - Frames per Second

GIT - Global Information Technologies

GM - General Motors

GOP - Group of Pictures

I.DRR - Integrated Driver Response Research

IIHS - Insurance Institute for Highway Safety

LED - Light Emitting Diode

NCAP - New Car Assessment Program

NHTSA - National Highway Traffic Safety Administration

OLED - Organic Light Emitting Diode

PCS - Pre-Collision System

RGB - Red, Green, and Blue

RPM - Revolutions per Minute

SDM - Sensing and Diagnostic Module

sUAS - Small Unmanned Aerial System

SUV - Sport Utility Vehicle

TCR - Traffic Collision Report

TD - Twilight Distance

TTC - Time to Collision

TTI - Time to Intersection

UAV - Unmanned Aerial Vehicle

US - United States

USGS - United States Geological Survey

VCH - Vehicle Control History

VD - Visibility Data

VIN - Vehicle Identification Number

VL - Visibility Level

VSP - Vehicle Sound for Pedestrians

VSS - Vehicle Speed Sensor

WREX 2000 - World Reconstruction Exposition in 2000

Chapter 01

Legacy Methods in Pedestrian Accident Reconstruction

We now live, and reconstruct car crashes, in a surveillance society. Gone is the time when accident reconstruction was focused purely on physical evidence and mechanical phenomenon—tire marks, gouges, vehicle deformation, broken glass, debris, and the like. Physical evidence is still relevant and useful, of course, but few crashes occur anymore without also leaving a trail of electronic data or video evidence from event data recorders (EDRs), cameras, and driver assistance systems. Accident reconstruction still relies on principles of physics and empirical data, but now is supplemented with electronic, video, and audio evidence. In many instances, an abundance of evidence can yield a fuller picture of how and why a crash occurred. But these new types of evidence give rise to new questions, and new and additional analysis methods and assumptions are needed for this new era.

Each of these newer sources of evidence has its idiosyncrasies and limitations. These need to be understood and accounted for by an accident reconstructionist attempting to integrate them into a coherent and reasonably complete picture of what occurred. As an example, consider a potential limitation introduced by the frame rate of a video. If a video consists of 4 fps (some older DriveCam videos have this frame rate), for instance, the interval of time between each frame will be approximately 250 ms. For an impact that might only last for 100 or 150 ms, it is possible that none of the frames will capture the impact itself.

In addition, there will always be what the camera sees and what it does not. A dash- or windshield-mounted camera will see out the front of the vehicle but usually will not capture what is happening behind or at the side of the vehicle. Critical aspects of the event may occur outside the view of the camera.

Other books exist that describe methods for reconstructing pedestrian collisions. Those books are still relevant and useful, but they were written prior to the ubiquitousness of electronic, video, and audio evidence. This book describes accident reconstruction methods for this new era, describing newer methods and considerations that are now needed to reconstruct a pedestrian collision. Each phase of the crash will be discussed—the human factors, pre-impact braking/skidding, impact, and post-impact phases. Methods for integrating event data and video evidence with legacy accident reconstruction methods will be introduced. This first chapter reviews the legacy methods. This includes methods for evaluating how a pedestrian interacts with a vehicle and the application of both theoretical and empirical models. These legacy methods rely primarily on physical and testimonial evidence. The physical evidence from a pedestrian collision may include: (1) the rest positions of the involved vehicle and pedestrian; (2) the rest positions of the pedestrian's clothing and accessories (such as shoes, hats, or bags); (3) tire marks from the involved vehicle; (4) damage, contact marks, or material transfer to the vehicle; (5) debris from the vehicle; (6) injury locations and types on the pedestrian; and/or (7) a scuff on the ground at the location of impact from the pedestrian's shoe(s). Documenting this evidence

may involve physically visiting the crash site or physically inspecting the involved vehicle, but could also rely on aerial imagery [1.1, 1.2], publicly available lidar measurement data [1.3, 1.4], or photogrammetric analysis. Thus, the methods that prior books on pedestrian accident reconstruction discuss will mostly be covered in this first chapter.

Subsequent chapters discuss newer vehicle systems relevant to pedestrian crash reconstruction, new types of evidence, and methods for integrating these systems and evidence into the reconstruction process. Chapter 2 discusses emergency-level braking for newer-model vehicles. The deceleration levels achievable with newer vehicles are higher than with older vehicles. In addition, antilock brake systems (ABSs) on these vehicles often prevent skid marks from being deposited, and brake assist systems potentially result in drivers utilizing more of the deceleration capabilities of the vehicle than they would have in the past. The proliferation of automatic emergency braking (AEB) and other driver assistance systems on newer vehicles also introduces a new wrinkle for accident reconstruction. The vehicle itself may now be capable of responding, and so accident reconstructionists will need to consider the degree to which they can distinguish the response of a driver from the response of the vehicle. This could be an important detail in a reconstruction since intervention by the vehicle could indicate a lack of response, or an untimely response, by the driver. AEB is typically paired with forward collision warning (FCW) such that the first intervention by a vehicle will be a visual, audible, and/or haptic warning that aims to prompt driver intervention to avoid a collision.

Chapter 3 discusses simulations of pedestrian collisions. This chapter brings focused attention to multibody simulation, where the interaction between the vehicle and the pedestrian is simulated in a detailed fashion. Use cases for such simulations are discussed, as are the strengths and limitations of this type of analysis. Chapter 4 discusses video analysis. Chapter 5 discusses data from vehicle EDRs in pedestrian collisions. Chapter 6 discusses the daytime and nighttime visibility of pedestrians and methods that can be used to assess when a driver could have detected a pedestrian.

Throughout the book, examples are drawn from real-world collisions and reconstructions carried out by the authors. When developing analysis methods with staged collision data, or with collisions captured on video, the answers to some of the relevant questions are known, or at least partially known. For many real-world collisions, though, there is no answer at the back of the book that we can reference. We can apply the best available analysis techniques and determine what is most probable. But often, there is disagreement between two experts examining the same crash and the same evidence. As you read the examples and case studies in this book, there is no need for you to accept the conclusions reached. The purpose of these case studies is not to demonstrate that we were right in our reconstructions or that another expert was wrong. Instead, the purpose is to illustrate the real-world applications of the techniques covered. In fact, that's how we have focused these case studies—on the illustration of

analysis methods. For the most part, the ultimate opinions reached for the underlying legal cases are left out.

When the reader arrives at the end of this book, they will be equipped with valid methods to evaluate the physical, video, audio, and testimonial evidence to answer questions such as the following: (1) Where on the road did the collision occur? (2) What was the impact speed of the vehicle? (3) How fast was the pedestrian moving at the time of the collision? Were they walking or running? What was their body posture? (4) Were the descriptions of the crash given by witnesses and involved parties accurate? (5) Were there visibility obstructions that contributed to the crash? (6) Was the pedestrian visible in time for the driver to have responded to avoid the collision? (7) What actions would have been needed by the driver to avoid the pedestrian? (8) What could the pedestrian have done to avoid the collision? (9) What were the lighting conditions at the time of the collision, and did those lighting conditions play a role in the crash?

Staged Collisions with Pedestrian Dummies

Beginning in the 1960s and continuing to the present day, the technical literature has reported numerous staged collisions between vehicles and pedestrian dummies. These staged collisions provide empirical data for reconstructing real-world pedestrian collisions, and they illustrate the types of interactions that can occur between vehicles and pedestrians. They provide data for the development of both theoretical and empirical models relating the pedestrian throw distance to the vehicle impact speed. They demonstrate the accelerations experienced by a pedestrian when struck by a vehicle or when striking the ground. They show the physical evidence present after a pedestrian collision and assist with the interpretation of that evidence.

As an example, Severy and Brink [1.5] reported a series of 12 staged collisions conducted between May and October of 1963 in which passenger cars struck pedestrian dummies. Multiple dummies were struck in each test— between the 12 tests, 38 dummies were impacted. The purpose was to study the influence of pedestrian size, posture, impact location on the vehicle, vehicle type, vehicle size, and vehicle speed on collision dynamics. Vehicle collision speeds varied between 10 and 40 mph. Three vehicle types were utilized: a sports car, a passenger car, and a truck. The sizes of the dummies varied: an adult (72 in., 200 lb), a 6-year-old (46 in., 48 lb), a 3-year-old (36 in., 36 lb), and a toddler (29 in., 32 lb). The majority of the tests were run with the medium-sized passenger car; a single test was run with the sports car and a single test with the truck. The location on the front of the vehicle where the pedestrians were struck was varied, as was the level of braking (none, moderate, panic). The dummies were dressed in "leisure-type clothing fitted to allow torn and abraded areas to be associated with the body area impacted. The dummies and their clothing were carefully inspected and photographed after each experiment to assure proper identification of injury patterns...glasses, shoes, hats, and similar objects were worn and carried by these simulated pedestrians to learn more about inertial action and displacements for

such objects." After each test, "both the car and the dummies were subjected to careful post-crash scrutiny to discover or refine investigative techniques of identification. Paint flecks on clothing, fabric embossment on car paint, as well as identification of shoe scuffs and head impacts with pavement, are examples of post-crash observations."

These authors observed that "the distances pedestrians are propelled by impacts from passenger vehicles increase exponentially with equal increases in impact speed." They also observed that, "for the same speed, impacts near the center of the vehicle front tend to project the pedestrian greater distances than impacts close to the right or left sides of the front bumper. Additionally, impacts occurring with the brakes 'off' propel the pedestrian greater distances than when impacts at the same speed occur with brakes applied at impact." For each test, an evidence diagram was presented that showed the post-test rest position of the vehicle, the pedestrian dummies, and various objects initially worn or held by the dummies (hats, glasses, and briefcases, for example). An example of one of these evidence diagrams is included in **Figure 1.1**. As this diagram shows, these authors created a polar grid for their tests to aid in their documentation of the rest positions of the vehicle, the pedestrian, and the debris. These tests reported by Severy and Brink are an example of the type of tests that have been reported in the literature. Many additional test series have been reported, with similar setup and intent, and many of these will be referenced in the discussion that follows.

Figure 1.1 Evidence from Severy and Brink, Test 67 (30 mph, brakes on).

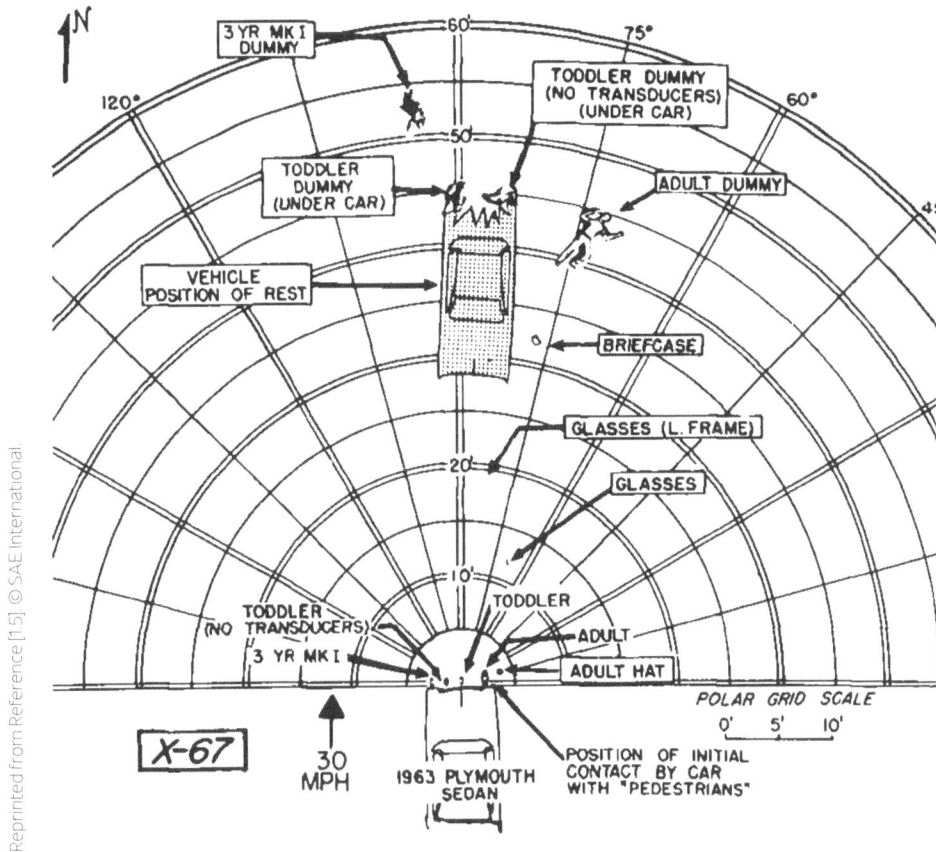

As another example, a series of staged collisions was conducted by the authors of this book. These tests involved a 2007 Chevrolet Malibu LS impacting a full-sized Rescue Randy pedestrian dummy. The Malibu, which was equipped with ABS, had an empty weight of approximately 3100 lb. In setting up the vehicle for testing, the driver and passenger airbags were removed. The vehicle was outfitted with a Racelogic Video VBOX Pro data logger. The VBOX recorded the vehicle speed and acceleration at 10 Hz, synchronized video from two cameras at 30 fps, and stereo audio. The cameras were mounted on the interior of the vehicle, one on the rear window facing forward and the other on the rear passenger side window facing the driver.

The dummy was a Nasco Healthcare "Rugged Rescue Randy" with a manufacturer-specified weight of 165 lb. During the testing, the dummy was outfitted with clothing and accessories, and the as-tested weight was 172 lb. The stature of the dummy was 5 ft, 9 in. The dummy was disassembled, and each body part was weighed. Through this process, the center of gravity (CG) of the dummy in an upright standing configuration was determined to be 42.9 in. above the ground. To hold the test dummy in place prior to the testing, a test rig was constructed from aluminum square tubing. The fixture was approximately 8 ft high and included a 12-ft-wide interior area for the vehicle to drive through. A mounting bracket was designed so that the pedestrian dummy could be moved laterally relative to the center of the rig for offset impacts. The test rig, test vehicle, and dummy are depicted in **Figure 1.2**.

Figure 1.2 Setup of pedestrian impact testing.

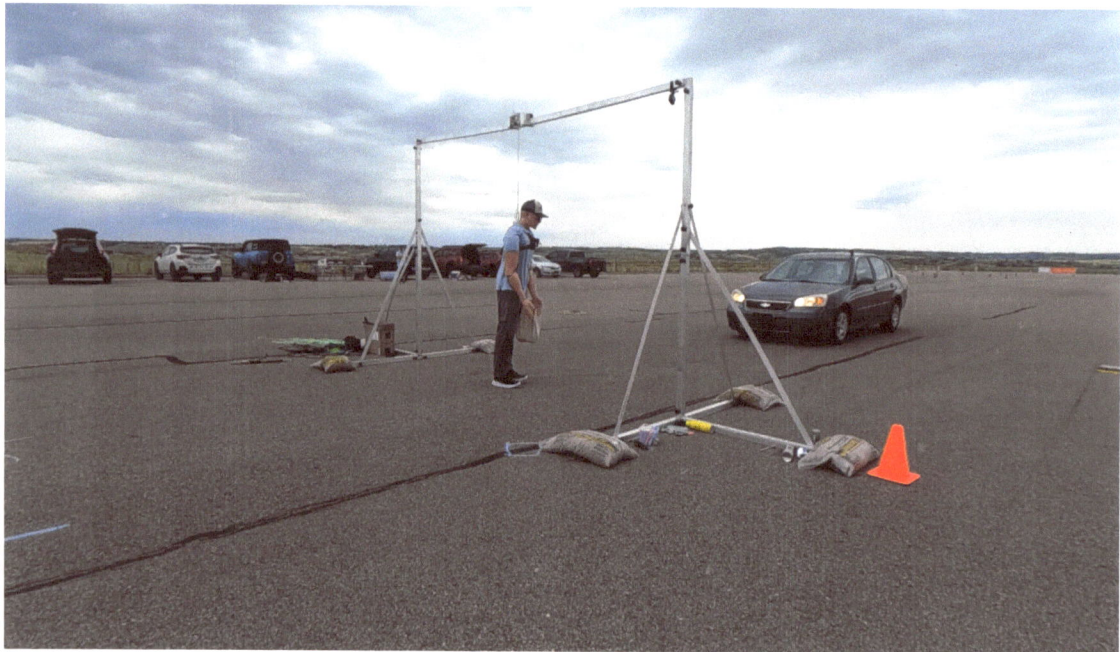

Four tests were conducted, two of which will be reviewed here (Test 2 and Test 3). In each test, the test vehicle was driven into the impact with aggressive braking applied prior to the impact. A frame of video from Test 2 is included in **Figure 1.3**, and a frame of video from Test 3 is included in **Figure 1.4**. In Test 2, the dummy was placed in the center of the test fixture to align with the center of the test vehicle. The driver accelerated the vehicle to a speed of 49 mph and then applied the brakes aggressively. At contact with the dummy, the vehicle was traveling at approximately 40.1 mph. The dummy contacted the front bumper and grille of the vehicle and then wrapped onto the hood and windshield in a location near the longitudinal centerline of the vehicle with a slight offset to the passenger side. An evidence diagram, developed using an orthomosaic image created with Pix4D mapper from the aerial imagery taken after the test, is included in **Figure 1.5**. As this diagram shows, the vehicle traveled 57.1 ft (17.4 m) after first contacting the dummy, and the dummy traveled 80.9 ft (24.7 m).

Figure 1.3 Test 2, collision between dummy and windshield, wrap.

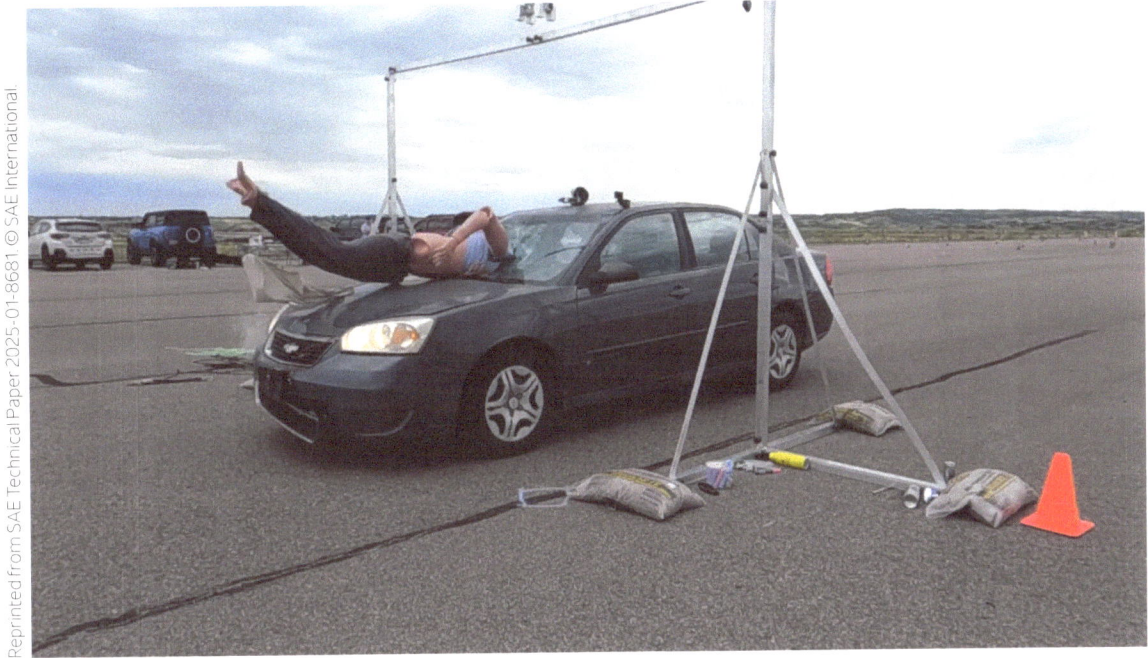

Figure 1.4 Test 3, collision between dummy and A-pillar, fender vault.

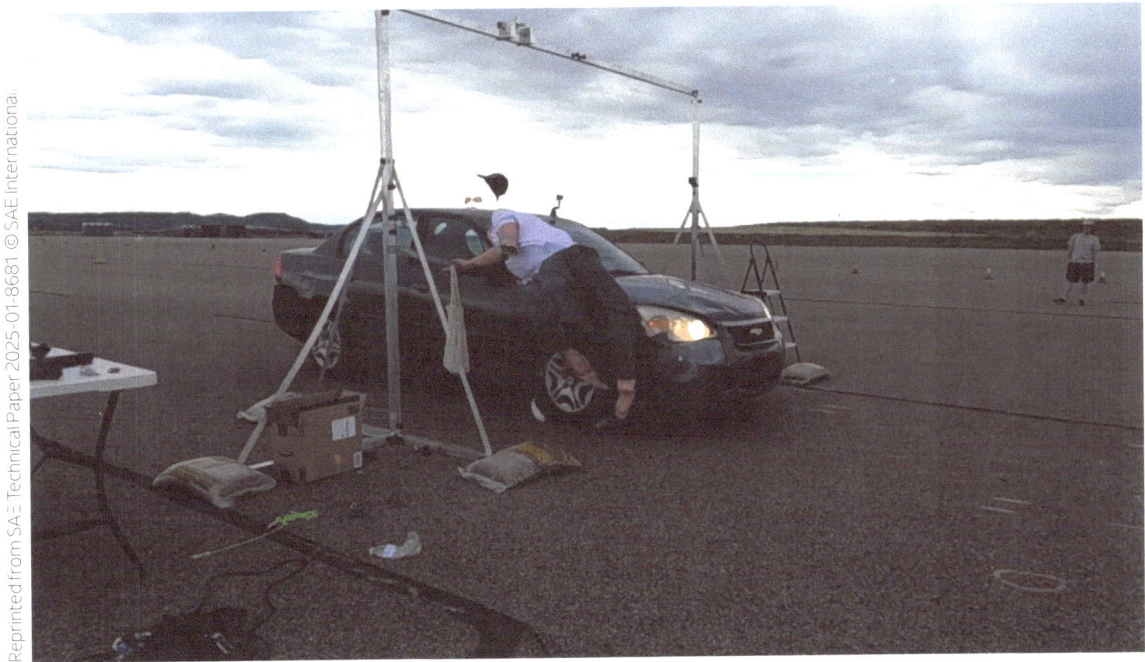

Figure 1.5 Test 2, evidence diagram.

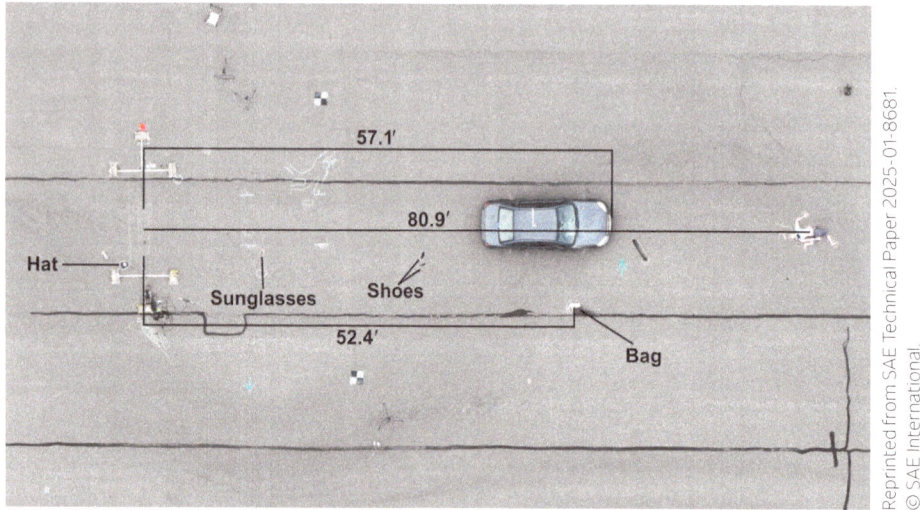

In Test 3, the dummy was offset to the side of the test fixture to align with the passenger side corner of the test vehicle. The driver accelerated the vehicle to a speed of 52 mph and then applied the brakes aggressively. At contact with the dummy, the vehicle was traveling at approximately 42.3 mph. The dummy was contacted by the front passenger side bumper and then wrapped onto the passenger side fender. The dummy's head contacted the A-pillar and windshield before contacting the roof of the vehicle. From the first contact to rest, the Malibu in this test traveled approximately 69.2 ft (21.1 m), and the test dummy traveled approximately 64.2 ft (19.6 m) along the longitudinal axis of the vehicle and approximately 23.6 ft (7.2 m) laterally to the vehicle for a total distance of 68.4 ft (20.8 m). An evidence diagram for this test is included in **Figure 1.6**.

Figure 1.6 Test 3, evidence diagram.

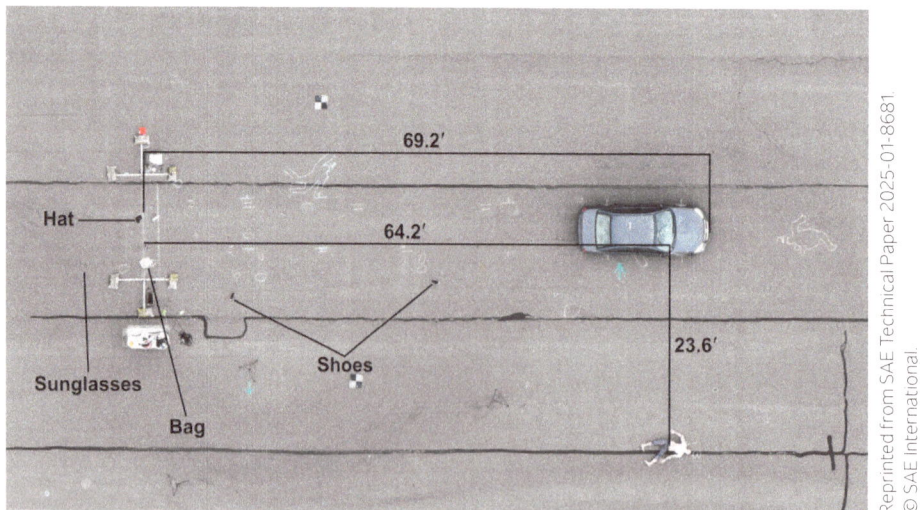

In both tests, the dummy was holding a bag. In Test 2, the bag was positioned such that it would be struck by the vehicle. In Test 3, the bag was positioned so that it would not be struck by the vehicle. In Test 3, the bag came to rest close to the point of impact, whereas in Test 2 the bag was propelled approximately 53 ft down the test surface. This illustrates the difficulty in utilizing items carried by a pedestrian to determine the location of the impact. The rest position of these items depends on how significantly they are engaged by the striking vehicle and on how significantly they are entangled with the pedestrian. The reconstructionist may not have any way to determine the degree of engagement experienced by such items and therefore will need to proceed with caution when attempting to use their final resting locations to determine the location of the impact. In both of these tests, the dummy's hat came to rest close to the area of impact. This is not unusual, but it is also important to state that it is not always the case. In Test 65 in the Severy and Brink study [1.5], the dummy's hat traveled 27 ft from the area of the impact. Also interesting is that in Test 2, the hat came to rest upstream of the area of impact, and so did the sunglasses in Test 3.

Roadway Evidence from Pedestrian Crashes

When reconstructing a pedestrian collision, there may be a question about where on the road the collision occurred. Sometimes this will be apparent from what the involved driver or witnesses say. If the collision is unwitnessed or the driver does not recognize the presence of the pedestrian prior to the collision, then physical evidence might illuminate the location of the collision. One possibility is a slight deviation in a skid mark that marks the location of impact. Such a deviation is likely to be small and might not be observed or documented by on-scene investigators. In addition, the presence of ABS on most modern vehicles makes this evidence less likely to be present. Another possibility is the transfer of material from the pedestrian's shoe onto the road surface. Fricke [1.6, 1.7] noted that on-scene investigators should "take special care to look for signs of the first contact positions…One of the best indicators of first contact is a mark from a shoe." Unfortunately, shoe scuffs are rarely identified by on-scene investigators. **Figure 1.7** is a photograph showing an example of a shoe scuff that was documented, along with the corresponding damage to the pedestrian's shoe. Speaking anecdotally, this shoe scuff is rare in how clear and visible it is. Often, shoe scuffs do not exhibit such contrast with the underlying road surface and are much more difficult to identify.

Figure 1.7 Example of a shoe scuff and associated shoe damage.

Shoe
scuff

(a)

(b)

Perhaps another possibility for determining the location of the collision is to infer it from the location of vehicle debris or the resting position of an item the pedestrian was carrying. However, the reconstructionist should proceed with caution when attempting to draw conclusions from such evidence. According to Fricke [1.7], "vehicle debris can give a clue to the general area of a collision. However, debris from a car can be a considerable distance from the first contact position…hats, canes, glasses and other things can be found some distance from the first contact positions." Collins [1.8] stated that "the point of impact can often be determined either by debris from the vehicle (primarily dirt dislodged from under the fender or broken headlight glass), or by the location on the pavement of clothing or other personal effects of the pedestrian. Hats, handkerchiefs, and other light objects do not travel far. Heavier objects, such as shoes, purses, and eyeglasses, can be used to indicate the limit of the area of impact." Unfortunately, not many pedestrians these days are carrying handkerchiefs. Severy and Brink [1.5] observed that "items of apparel and objects being carried such as briefcases, purses, and so forth, generally are not propelled as far as the pedestrian, in part because they tend to become dislodged from the pedestrian before he is fully accelerated and, in part, because these objects are generally of lesser density than the pedestrian. Hard objects such as canes, directly impacted by the car may be projected further than the pedestrian. Soft objects, such as caps securely attached may remain affixed to the pedestrian until dislodged as he impacts or scuffs to a stop; in such instances these objects may travel distances as far or further than the pedestrian."

The rest locations of the dummies and their hats and other objects from additional Severy and Brink's tests are included next. **Figure 1.8** depicts the rest positions of the pedestrian dummies along with a hat and a briefcase for a test involving a braked passenger car colliding with four pedestrian dummies at a speed of 20 mph (Test 65). In this test, the hat worn by the adult dummy came to rest 27 ft from the collision, traveling further after the collision than the adult dummy. **Figure 1.1** depicts the rest positions of the pedestrian dummies along with a hat, a briefcase, and a pair of glasses for a test involving a braked passenger car colliding with four pedestrian dummies at a speed of 30 mph. In this test, the hat worn by the adult dummy came to rest within a few feet of the collision. The briefcase traveled approximately 32 ft from the area of the collision, and pieces of the glasses came to rest between 10 and 20 ft from the collision. **Figure 1.9** depicts the rest positions of the pedestrian dummies along with a hat worn by the adult dummy for a test involving a braked passenger car colliding with three pedestrian dummies at a speed of 40 mph. In this test, the hat worn by the adult dummy came to rest within a couple of feet of the collision. **Figure 1.10** depicts the rest positions of the pedestrian dummies along with a hat worn by the adult dummy for a test involving a braked 1956 Corvette colliding with three pedestrian dummies at a speed of 30 mph. In this test, the hat worn by the adult dummy came to rest within 5 ft of the collision. **Figure 1.11** depicts the rest positions of the pedestrian dummies along with a hat worn by the adult dummy for a test involving a braked 1963 Ford 1½ ton truck colliding with three pedestrian dummies at a speed of 30 mph. In this test, the hat worn by the adult dummy came to rest approximately 30 ft from the collision.

Figure 1.8 Severy and Brink, Test 65, 20 mph, brakes on.

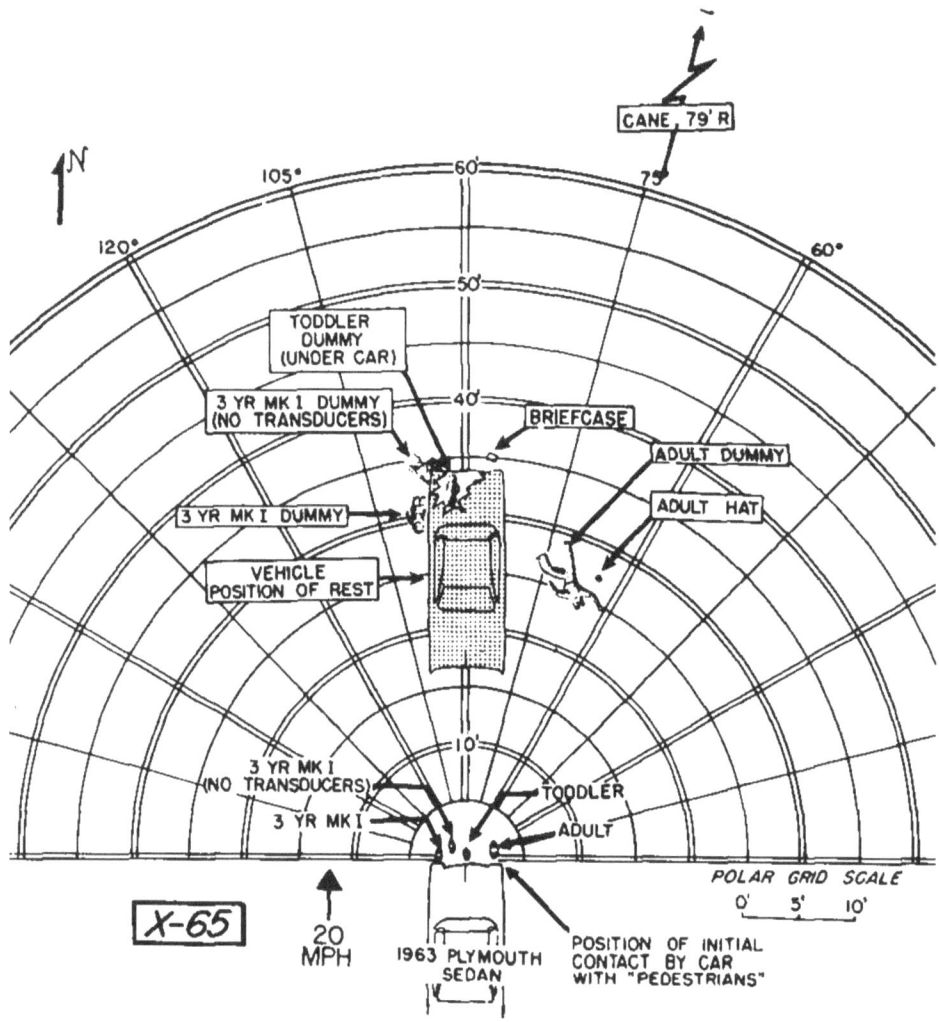

Figure 1.9 Severy and Brink, Test 70, 40 mph, brakes on.

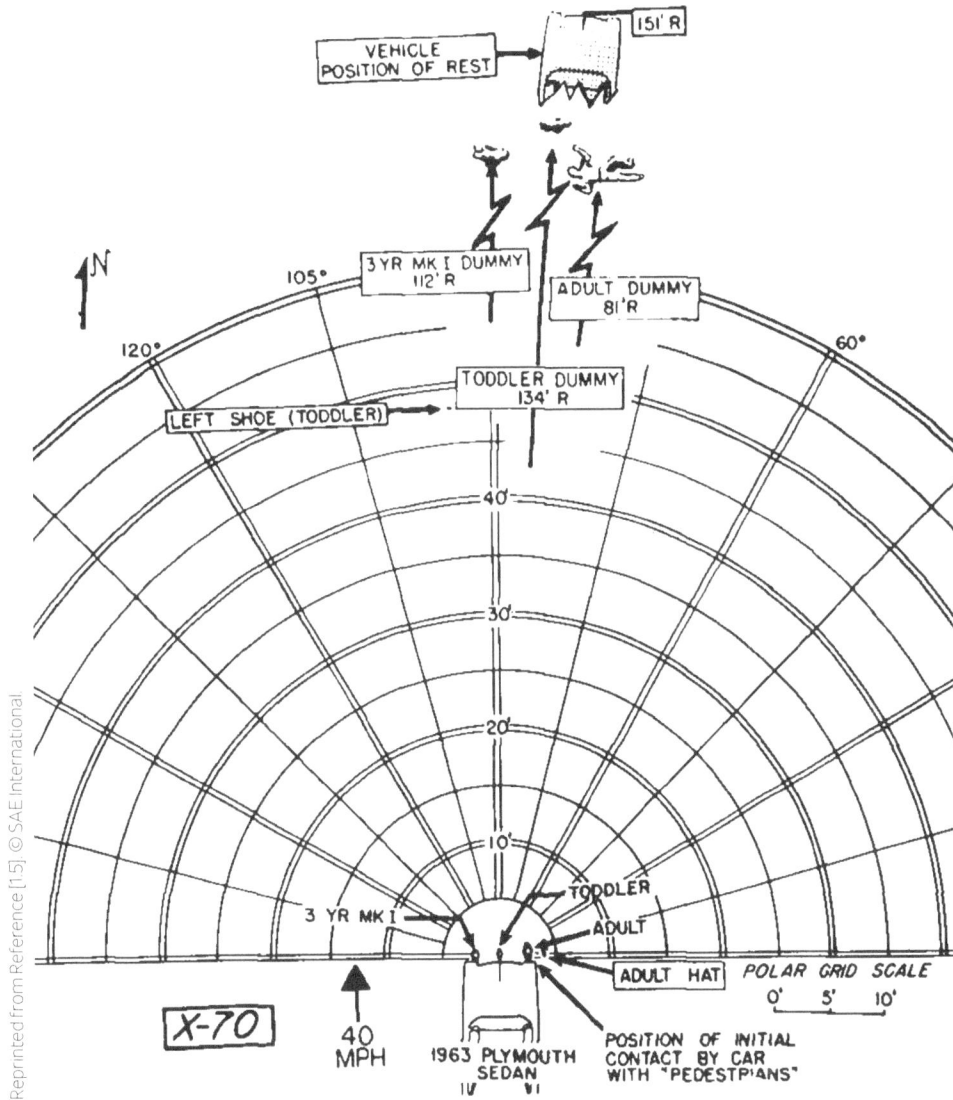

Figure 1.10 Severy and Brink, Test 74, 30 mph, brakes on, 1956 Corvette.

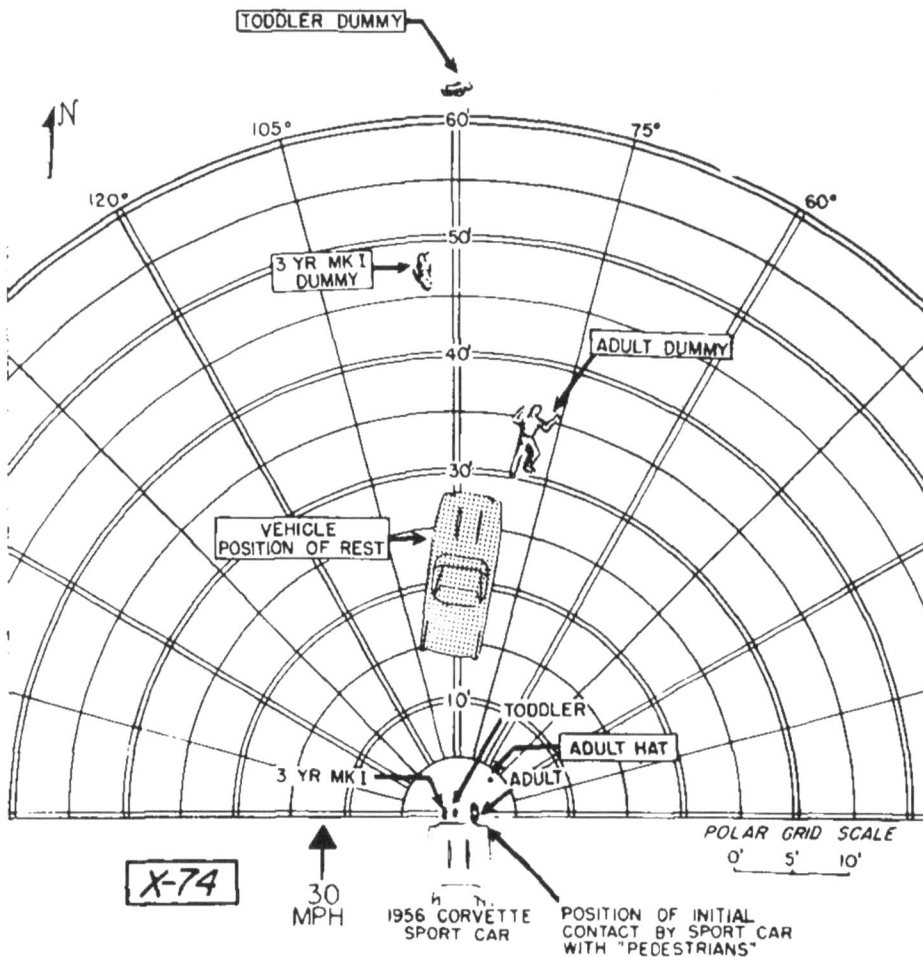

Figure 1.11 Severy and Brink, Test 75, 30 mph, brakes on, 1963 Ford 1½ ton truck.

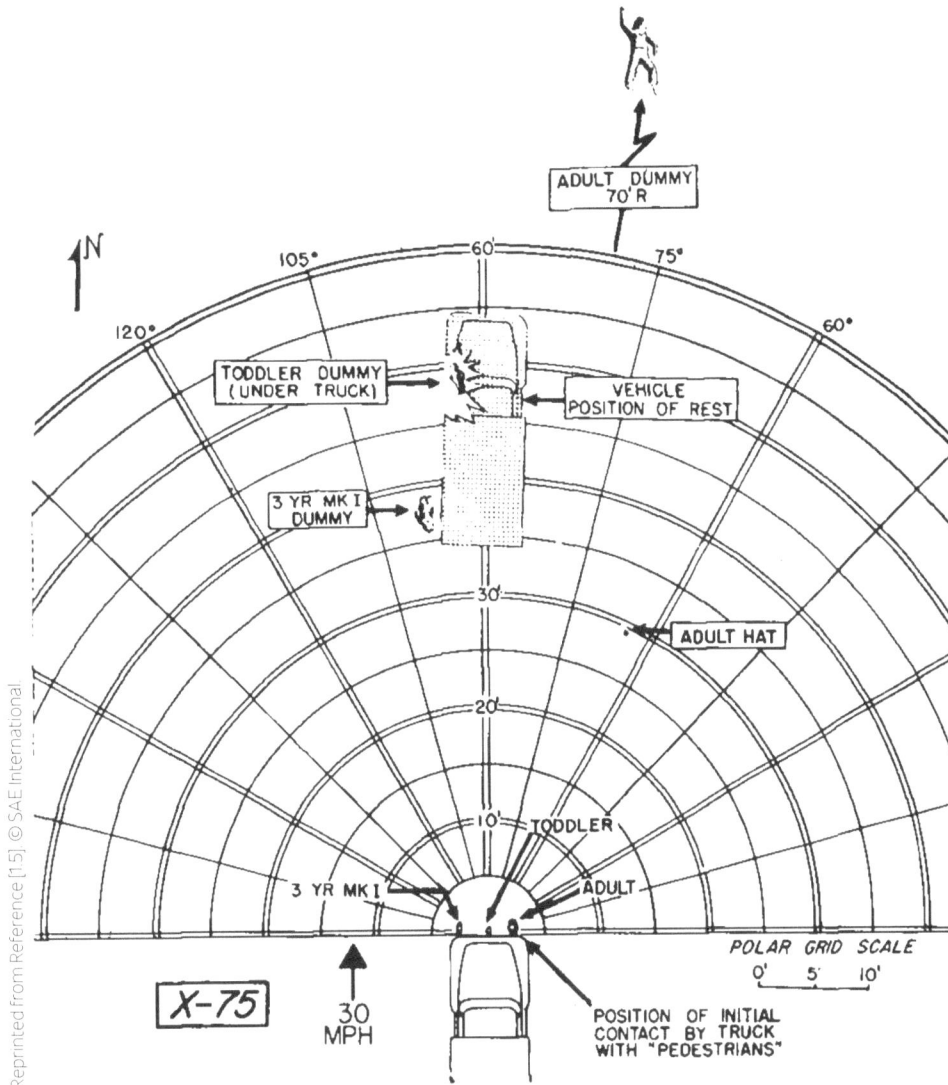

Prior literature seems to stop short of saying: "Don't use the resting locations of clothing items or carried items to determine the location of the impact with a pedestrian." And perhaps this is with good reason. Surely, it is *usually* the case that *most* of the debris, clothing, or carried items come to rest downstream of the area of impact.

What should be said decisively, though, is this: the resting locations of clothing items or carried items will not, without other supporting evidence, give you a precise impact location. Reconstructions will, of course, vary in the needed level of precision.

Pedestrian Impact and Trajectory Types

After a pedestrian collision, the damage to the vehicle, the location of the pedestrian's injuries, and the rest positions of the vehicle and the pedestrian will often reveal the type of collision that occurred. Understanding the nature of the collision is important for determining which theoretical or empirical models are appropriate for the analysis. If the pedestrian is struck by the front of the vehicle, then the motion of the pedestrian can often be characterized with one of the following categories: (1) a forward projection; (2) a wrap; (3) a fender vault; (4) a carry; or (5) a roof vault [1.9]. The following factors will influence which of these will occur: (1) the height of the pedestrian's center of mass compared to the height of the upper leading edge of the striking vehicle; (2) which portion of the vehicle strikes the pedestrian; (3) the speed of the vehicle; (4) whether or not the vehicle is being braked at the time of the collision; and (5) the speed of the pedestrian. In addition to these categories, a pedestrian can also be sideswiped or backed into or onto.

Forward Projection: If the upper leading edge of the vehicle is higher than the pedestrian's center of mass, the impact will result in a *forward projection*. This trajectory type is common with blunt and high-fronted vehicles, such as buses, trucks, sport utility vehicles (SUVs), and work vans or with shorter pedestrians. A forward projection results in the pedestrian's center of mass remaining ahead of the vehicle and a high percentage of the vehicle's impact velocity being imparted to the pedestrian. For impacts between pedestrians and SUVs, the pedestrian's upper torso and head may wrap onto the hood of the vehicle, but if the center of mass of the pedestrian remains in front of the leading edge of the vehicle, then the trajectory should be classified as a forward projection. **Figure 1.12** contains several illustrations of typical forward projections [1.9]. In each of these illustrations, a pedestrian is impacted by a vehicle with a front end high enough that the upper leading edge is above the pedestrian's CG and the pedestrian is projected out in front of the vehicle.

Figure 1.12 Forward projection examples [1.9].

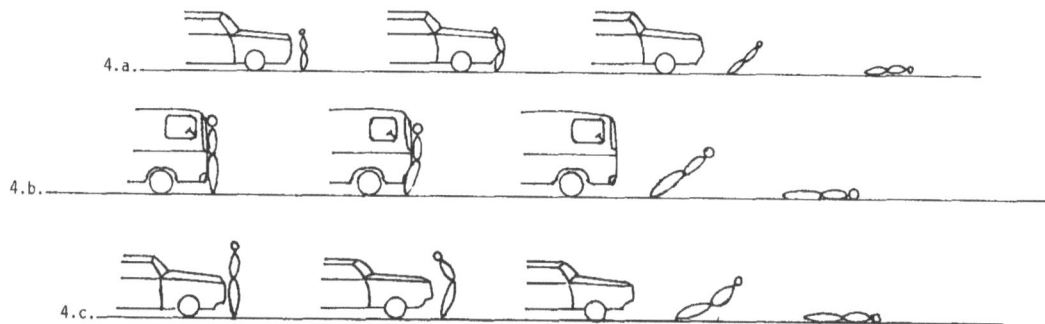

Figure 1.13 is a series of frames from a forward projection simulated with the PC-Crash multibody pedestrian model (ped to ground friction = 0.6, ped to vehicle friction = 0.2, restitution = 0.1). This model is helpful for demonstrating the influence of various parameters on pedestrian motion during a collision. This pedestrian model has been extensively validated and is covered in detail in Chapter 3. The multibody pedestrian consists of a series of rigid bodies representing the various body parts (the head, torso, and pelvis, for example) connected with joints. The user can specify the weight and height of the pedestrian, the initial position and orientation of each body part (within the constraints imposed by the joints), and the pedestrian's initial velocity. The software will sense the contact between the pedestrian's body parts and the vehicle or the ground in the simulation and will calculate the forces and resulting motion for the pedestrian. In the simulation depicted in **Figure 1.13**, the van is initially traveling at 35 mph with heavy braking applied. This vehicle has a flat front end,

and the leading edge of the hood is situated above the CG of the pedestrian. The pedestrian is struck on the right side while traveling at an initial speed of 3 mph with the left leg leading. The pedestrian's hips are initially situated at the center of the van. The frames in the figure are separated by 100 ms. As these frames show, the pedestrian is accelerated and reaches a common velocity with the van. The pedestrian's hips always remain ahead of the leading edge of the vehicle. In this simulation, the pedestrian's upper body wraps onto the hood of the van, and the pedestrian's head impacts the hood at approximately 66 ms into the simulation. Despite this wrapping onto the hood, the center of mass of the pedestrian remains ahead of the leading edge of the vehicle throughout the simulation, and so this remains a forward projection, rather than a wrap (the next trajectory type discussed). The pedestrian in this simulation is thrown a total distance of 83.2 ft—82.9 ft along the travel direction of the van and 7.5 ft laterally along the direction the pedestrian was initially walking.

Figure 1.13 Forward projection, van = 35 mph, heavy braking, pedestrian = 3 mph.

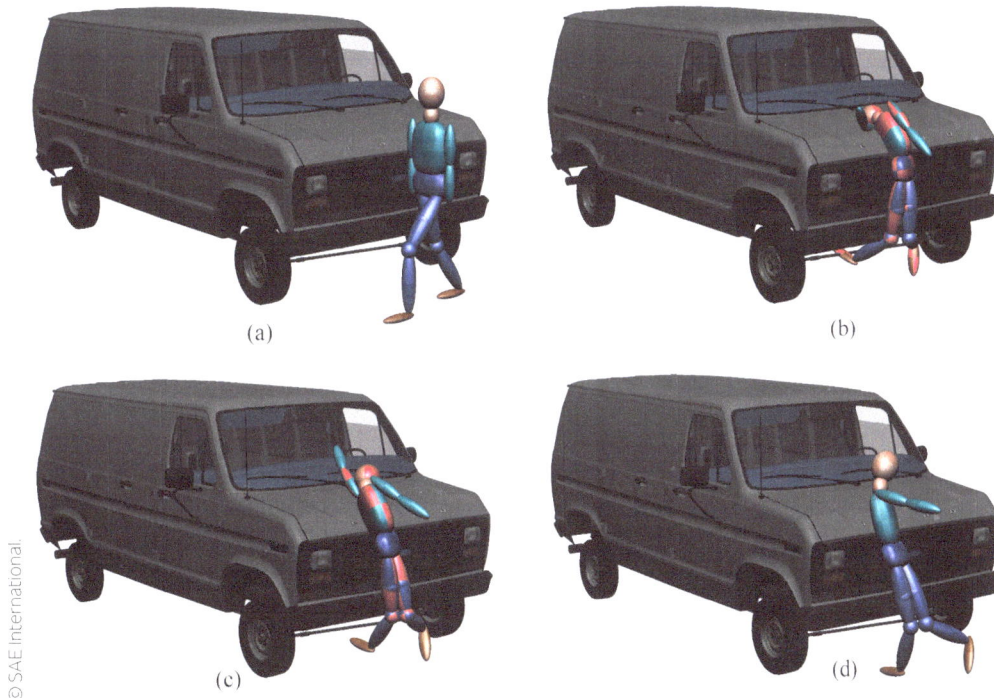

(a)   (b)

(c)   (d)

Another simulation identical to this one was run, with the exception that the pedestrian's initial speed was increased to 6 mph. The head impact in this second simulation occurred at 66 ms, identical to the prior simulation. The timing of this head impact was primarily driven by the speed of the striking vehicle, not the speed of the pedestrian. The location of the head impact on the hood was slightly different though, reflecting the velocity of the pedestrian relative to the vehicle. In this instance, the head impact was approximately 3 in. further away from the centerline of the vehicle than in the prior simulation. In the prior simulation, the center of the head had impacted the hood at approximately 3 in. toward the passenger's side from the centerline. In this simulation, the center of the head impacted the hood at approximately 6 in. away from the centerline toward the passenger's side of the vehicle. This difference in head impact locations could potentially assist a reconstructionist in determining the speed of the pedestrian, though being able to discern such a small difference in head impact location would depend on having precise documentation of vehicle damage. A sensitivity study related to other parameters of the model would also be needed, such as the pedestrian-to-vehicle friction coefficient. With the increase in the pedestrian speed to 6 mph, the lateral throw distance increased to 11.5 ft, reflecting the additional lateral velocity of the pedestrian. This lateral throw distance is another piece of evidence that could contribute to the determination of the pedestrian speed.

A later section of this chapter discusses empirical equations that relate the total throw distance of the pedestrian to the speed of the impacting vehicle. These equations have one input—the total throw distance—and one output—the vehicle impact speed. There are no similar equations for determining the speed of the pedestrian at impact, but simulations can potentially be used for this purpose, with the lateral throw distance being indicative of the pedestrian speed. This does, though, assume that the physical evidence on the vehicle is sufficient to determine which part of the vehicle struck the pedestrian, since the shape of the vehicle and the pedestrian's position along the front of the vehicle can also be influential in the lateral throw distance. To illustrate this, additional simulations were run, again with the pedestrian traveling at 6 mph. One simulation was conducted with the pedestrian being impacted on the passenger's side of the front end of the van. The lateral throw distance increased to 12.9 ft. Another was conducted with the pedestrian being impacted on the driver's side of the front. The lateral throw distance was reduced to 8.2 ft.

Wrap Trajectory: A *wrap* occurs when the pedestrian is struck by the front of the vehicle and the pedestrian's center of mass is higher than the leading edge of the striking vehicle. This trajectory is characterized by rearward movement of the pedestrian relative to the vehicle such that the pedestrian rotates onto the hood. Relative to the vehicle, the pedestrian's center of mass moves toward and behind the leading edge following first contact. This involves sliding and rotation of the pedestrian onto the hood and, often, secondary contacts of the pedestrian's hips, upper torso, or head with the hood, windshield, or windshield frame. This trajectory type imparts a smaller percentage of the initial vehicle velocity to the pedestrian than a forward projection does. Vehicle deceleration from braking is necessary to generate the separation that ultimately occurs between the pedestrian and the vehicle for this trajectory type [1.9]. When this braking is present, the pedestrian will come to rest ahead of the vehicle. If the vehicle is not decelerating at impact, the wrap is likely to become a carry or a roof vault. Wrap trajectories are common when the striking vehicle has a low leading edge (sedans and coupes). **Figure 1.14** illustrates the initial portions of a wrap trajectory [1.10]. As this figure shows, there are typically contacts between the vehicle and the pedestrian's leg, hip, shoulder, and head. The timing in this figure is illustrative; the actual timing can vary depending on the collision speed and the specific alignment between the pedestrian and the vehicle.

Figure 1.14 Example of a wrap.

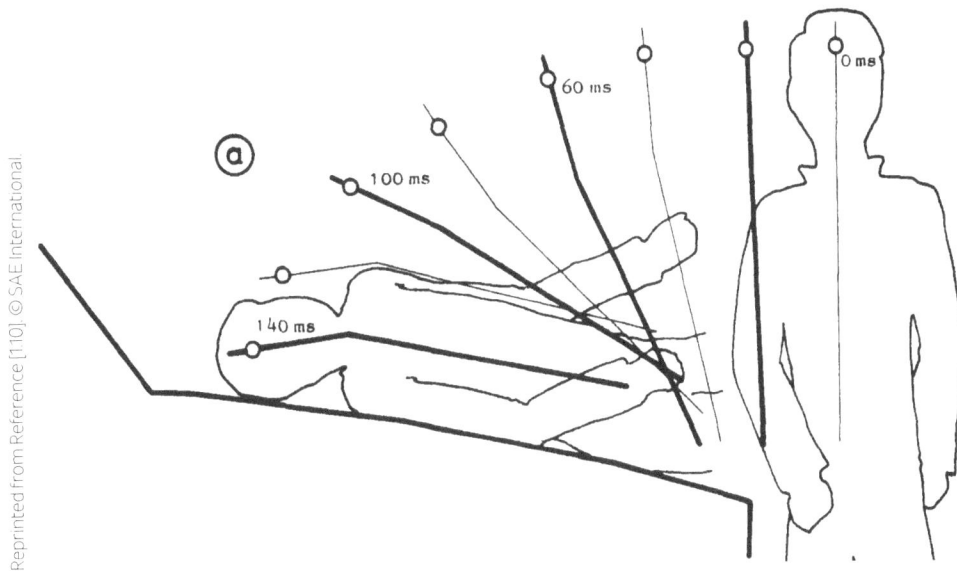

Reprinted from Reference [1.10] © SAE International.

Figure 1.15 is a series of frames from a wrap trajectory simulated in PC-Crash. In this simulation, the car is initially traveling at 40 mph with heavy braking applied. The pedestrian is struck on the right side while traveling at an initial speed of 3 mph with left leg leading. The pedestrian's hips are initially situated just to the passenger's side of center. The frames in the figure are separated by 30 ms. As these frames show, the trailing right leg is struck first and, as the pedestrian's hips begin moving onto the hood of the vehicle, the upper body is rotating in a clockwise fashion (viewed from above), exposing the front of the pedestrian's body to impact with the hood and windshield. The pedestrian's head strikes the base of the windshield at 96 ms into the simulation, just to the passenger's side of center. The pedestrian in this simulation was thrown a total distance of 75.2–74.5 ft along the travel direction of the car and 10.2 ft laterally along the direction the pedestrian was initially walking.

Figure 1.15 Wrap simulation, car = 40 mph, heavy braking, pedestrian = 3 mph.

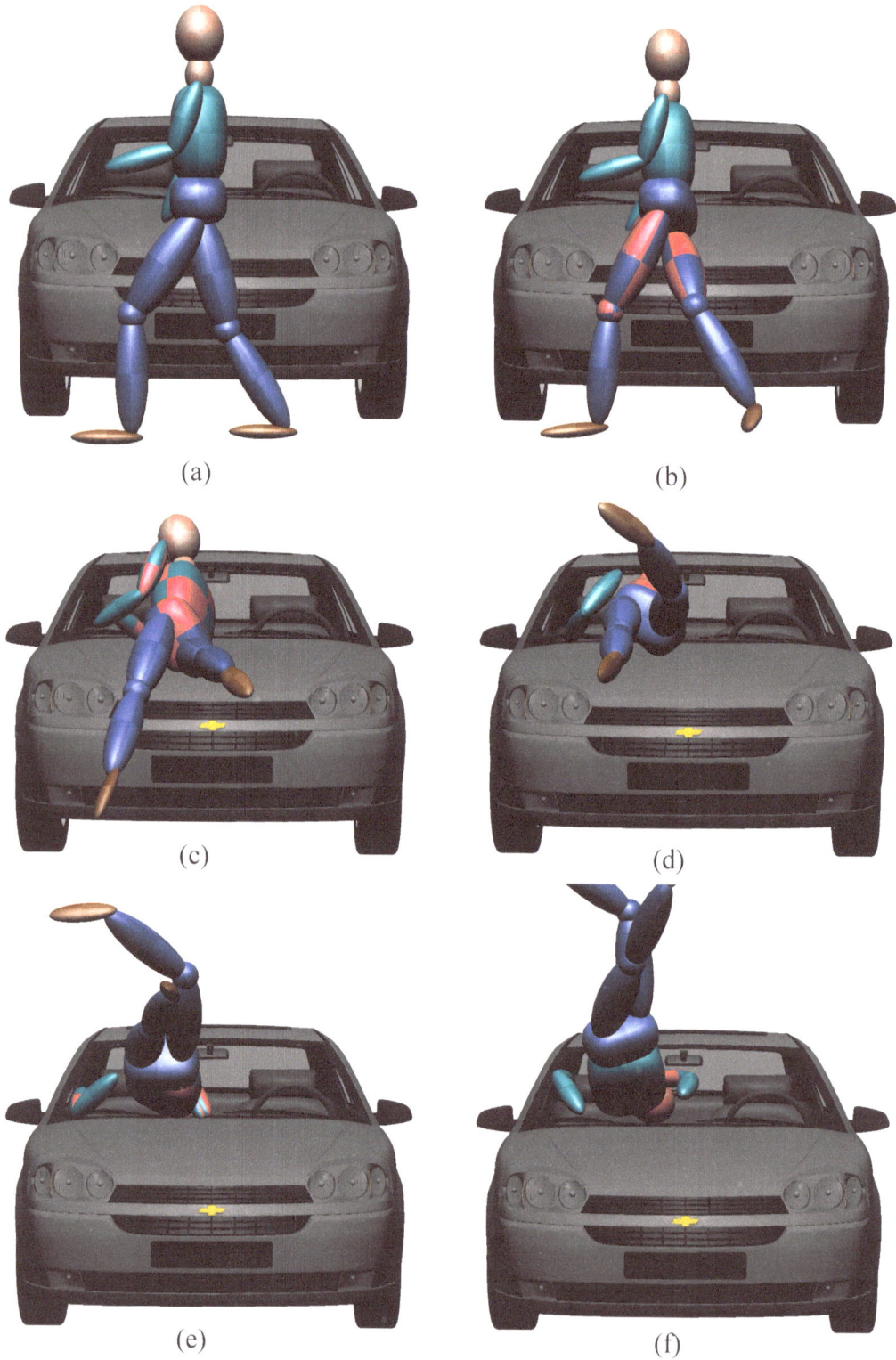

(a)

(b)

(c)

(d)

(e)

(f)

Another wrap trajectory was simulated in PC-Crash. This simulation was identical to the previous one with the exception that the pedestrian's initial speed was increased to 6 mph. With this increase in the pedestrian speed—a running or jogging pedestrian rather than a walking pedestrian—the head strike on the windshield occurred further toward the passenger's side of the vehicle. The difference in head strike location between the pedestrian speeds of 3 and 6 mph was approximately 5 in. The total throw distance of the pedestrian was 74.4 ft—less than a foot difference from the prior simulation. However, the lateral throw distance increased to 16.3 ft, reflecting the additional lateral velocity of the pedestrian.

Fender Vault: A *fender vault* is an incomplete wrap trajectory in that the pedestrian slides off the side of the vehicle and separates before the pedestrian is accelerated to the full speed they would have been by a complete wrap trajectory. This trajectory often occurs when the pedestrian is struck near one of the corners of the front of the vehicle, and the lateral velocity of the pedestrian generated by the curved front bumper contributes to the pedestrian sliding off the side of the vehicle. The speed of the pedestrian can also contribute. These trajectories can occur with braked or unbraked vehicles. For a fender vault trajectory, the rest position of the pedestrian will be either behind and to the side or next to the vehicle. **Figure 1.16** is a simulated illustration of a fender vault (ped to ground friction = 0.6, ped to vehicle friction = 0.2, restitution = 0.316). In this simulation, the vehicle was initially traveling at 40 mph and the pedestrian was traveling at 3 mph. The total throw distance of the pedestrian in this simulation was 49.4 ft, with a lateral component of 20.5 ft. In this simulation, the pedestrian's head struck the A-pillar of the

vehicle. When the pedestrian speed was increased to 6 mph, the pedestrian's head missed the A-pillar, and the total throw distance was reduced to 35.2 ft.

Carry: A *carry* can occur when the driver of a low-fronted striking vehicle is not braking at impact. This trajectory type occurs when the pedestrian and the vehicle maintain contact beyond what is typical of a wrap trajectory. The vehicle carries the pedestrian along the vehicle's path, usually until the vehicle driver brakes and the pedestrian then separates from the vehicle. **Figure 1.17** contains frames from a pedestrian carry simulated in PC-Crash. The vehicle in this simulation was initially traveling at 20 mph and the vehicle was not being braked at impact. The pedestrian was initially stationary. In this simulation, the pedestrian wraps onto the hood and then is carried by the vehicle until the driver of the striking vehicle begins braking aggressively, 1.5 sec after impact. If a pedestrian who is carried by the vehicle ultimately comes to rest out in front of the vehicle (which is what occurred in this simulation), then the physical evidence from a carry can look similar to a wrap trajectory. Distinguishing the two comes down to establishing when braking of the striking vehicle began. This can be established from event data from the striking vehicle's airbag control module (ACM), from skid marks, from evidence of where on the road the pedestrian struck the ground and began sliding, or from the striking driver's statement. Distinguishing a carry from a wrap is necessary for the reconstruction because the carry phase eliminates a simple relationship between the impact speed and the total throw distance. Empirical throw distance equations are appropriate for wraps but not for carries.

Figure 1.16 Fender vault simulation, car = 40 mph, pedestrian = 3 mph.

(a)

(b)

(c)

(d)

(e)

(f)

© SAE International

On the other hand, if a carry can be established, this can assist with the speed analysis when it is being conducted with simulation. In the simulation shown in **Figure 1.17**, the pedestrian rotates onto the windshield and is carried there. Increasing the impact speed of the vehicle in this simulation to 25 mph resulted in the pedestrian rotating onto the roof and being carried along on the roof. As a result of the heavy braking that ensued 1.5 sec into the simulation, the pedestrian still came to rest out ahead of the vehicle, but the impact speeds of 20 and 25 mph were distinguishable based on where on the vehicle the pedestrian reached a common velocity with the vehicle and was carried. In a

real-world case, there would likely be physical evidence of the pedestrian making it onto the roof of the vehicle (or not making it onto the roof of the vehicle). When the vehicle speed in this simulation was increased to 30 mph, the pedestrian never reached a common velocity with the vehicle and ended up at rest behind the vehicle. This is referred to as a roof vault, which is the next trajectory type discussed. This is a basic illustration for one vehicle shape, with one pedestrian gait position, and a single set of simulation inputs. The precise speed at which the trajectory type transitions from a carry to a roof vault will depend on the specifics of the case.

Figure 1.17 Carry simulation, vehicle = 20 mph.

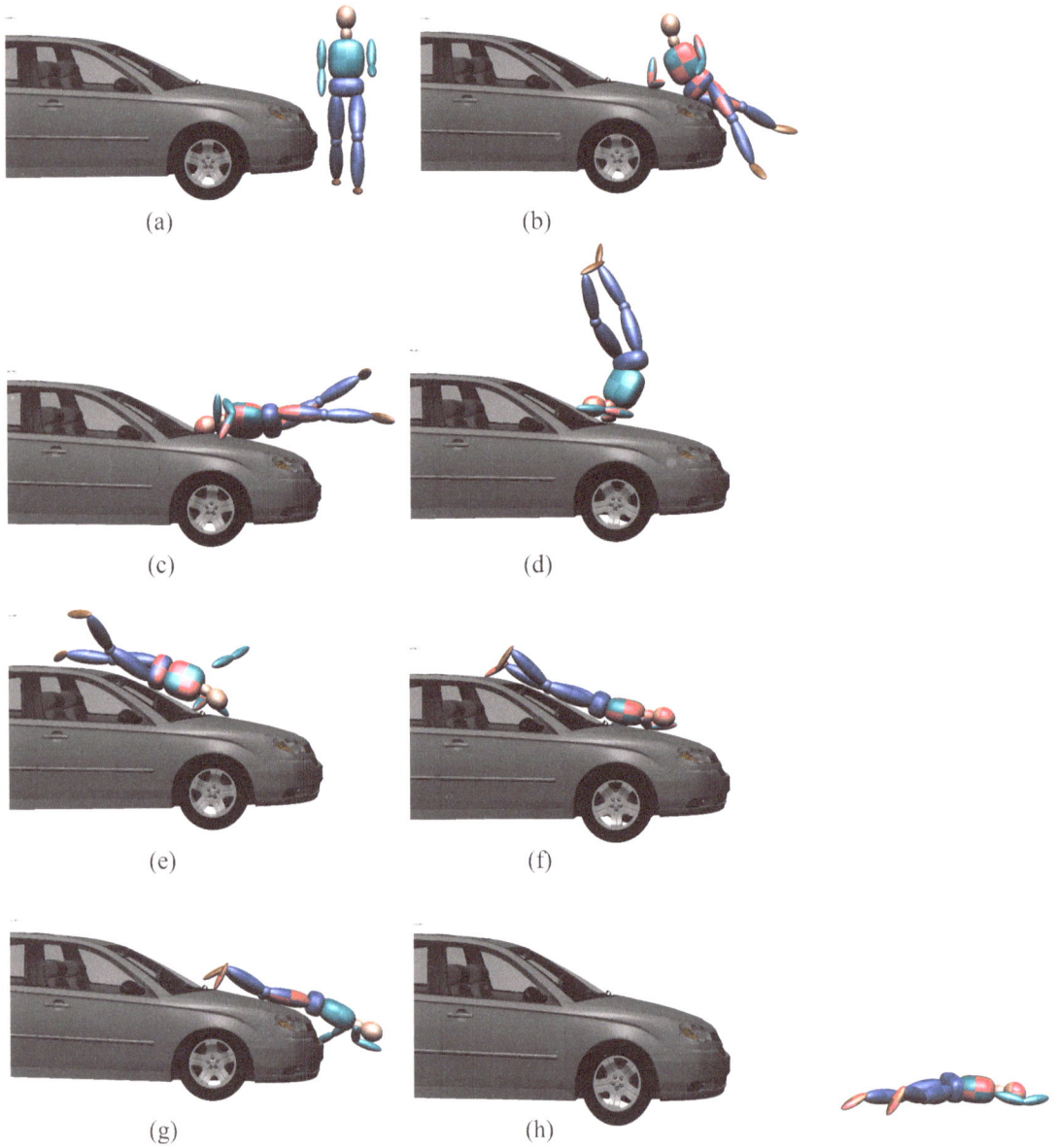

(a)

(b)

(c)

(d)

(e)

(f)

(g)

(h)

© SAE International.

Roof Vault: A *roof vault* occurs when a pedestrian is struck by the front of a vehicle that is not being braked and the pedestrian rotates over the top of the vehicle without ever reaching a common velocity with the vehicle. There will usually be some contact between the pedestrian and the hood, windshield, roof, rear window, or trunk lid of the vehicle, and these contact points can be used as a part of the reconstruction when using simulation. In a roof vault, the pedestrian

will come to rest behind the vehicle. The speed at which this will occur depends on the shape of the vehicle, the height of the pedestrian, the gait position, and likely other factors. Simulation can be an excellent approach for determining this for a specific case. An example is shown in **Figure 1.18**. In this simulation, the vehicle speed at impact was 30 mph and braking was not applied until 1.5 sec into the simulation.

Figure 1.18 Roof vault simulation, vehicle = 30 mph.

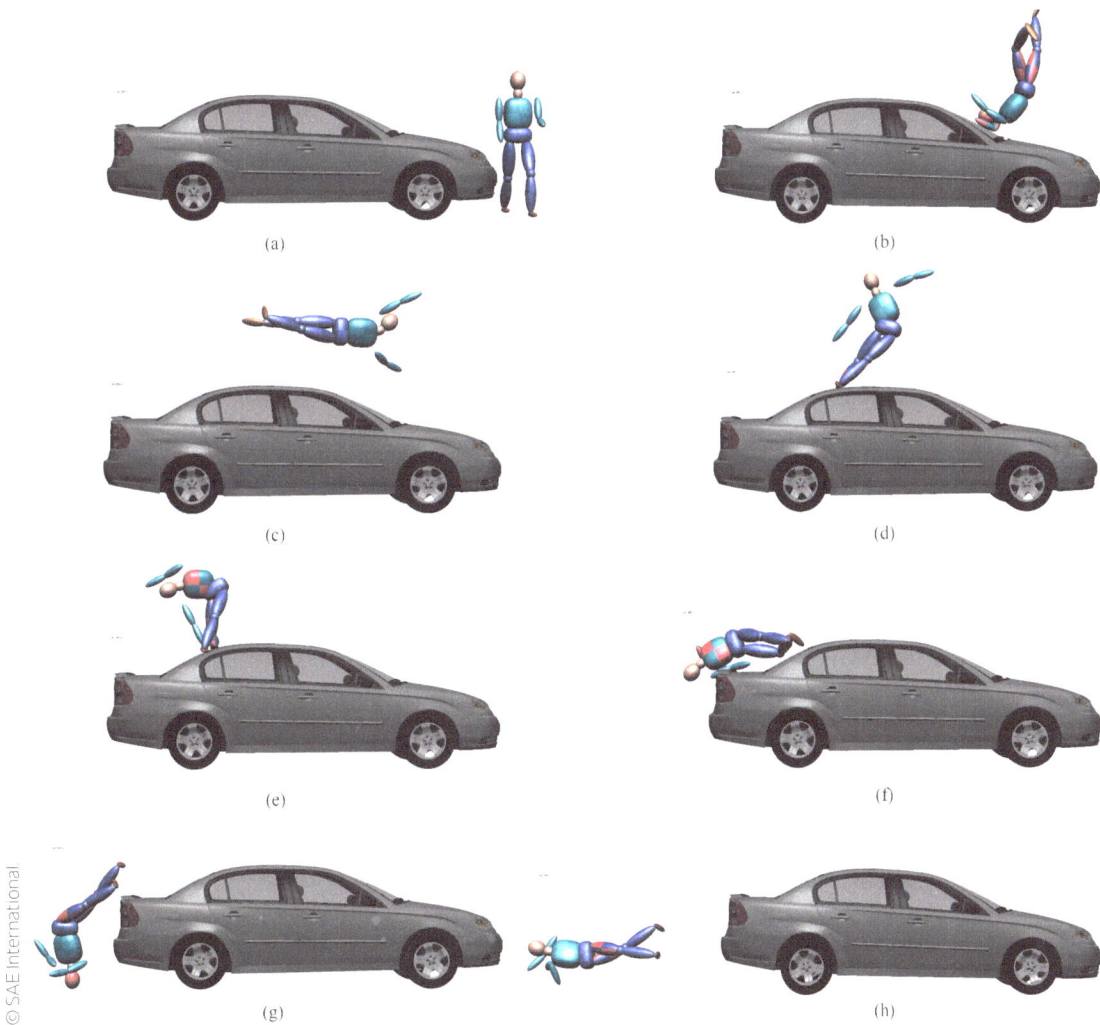

(a) (b) (c) (d) (e) (f) (g) (h)

© SAE International

For most of these trajectory types, the pedestrian will experience an airborne phase following the collision, then a landing phase, and a sliding/tumbling phase. The *throw distance* for any of these trajectory types is the total distance from the pedestrian's location at the first contact to the pedestrian's position of rest, including the impact, airborne, landing, and sliding/tumbling phases. For forward projection and wrap trajectories, empirical models have been developed relating the throw distance to the vehicle impact speed. These are discussed later in this chapter. Such empirical models are not applicable to other trajectory types, though Neale et al. [1.11] attempted to apply them to narrow overlap (fender vaults) and sideswipe impacts. This approach is critiqued later in this chapter.

Evidence on the Vehicle

Identifying contact locations of a pedestrian on a striking vehicle can assist with classifying the type of interaction that occurred (i.e., wrap, forward projection, fender vault, etc.) and therefore with selecting the correct method for analysis. As an example, **Figure 1.19** depicts a sedan that struck a pedestrian. The collision resulted in a wrap trajectory, and the interaction with the pedestrian's lower body damaged the hood, front fender, and headlight assembly at the passenger's side front corner. The pedestrian's upper body and head damaged the upper portion of the hood and the windshield and wiper. Hair from the pedestrian's head was lodged in the broken portion of the windshield. Similarly, **Figure 1.20** shows an instance of evidence markers being used to identify individual contact locations from a pedestrian on a Toyota Prius. **Figure 1.21** shows close-ups of some of these contact locations.

Figure 1.22 contains a series of video frames showing the dummy contacts with the vehicle during Test 2 conducted by the authors. These frames show the dummy contacting the vehicle in a manner consistent with a wrap trajectory, with the legs contacting the front bumper, grille, and leading edge of the hood, the hips contacting the hood, and the shoulders and head impacting the windshield. If the speed of the vehicle in this test had been lower, then the head impact with the windshield would have occurred at a lower position.

Figure 1.19 Pedestrian contact locations on a striking vehicle.

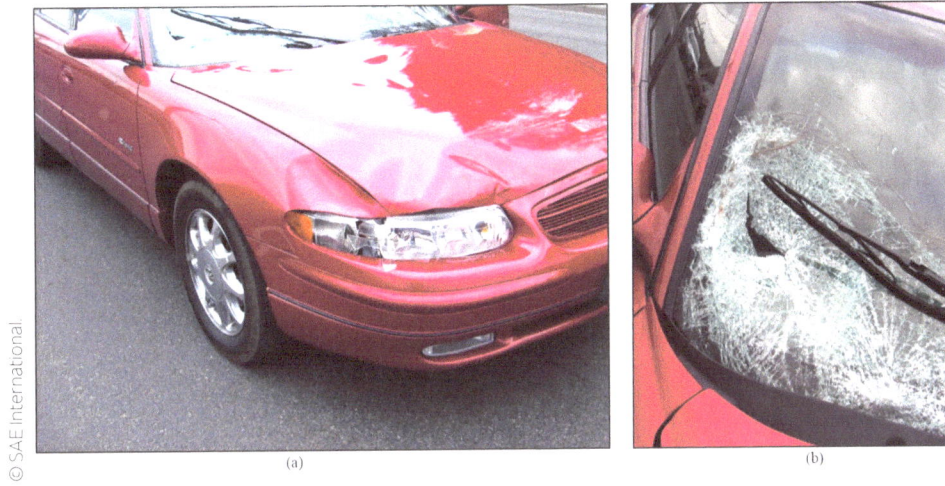

© SAE International.

(a)   (b)

Figure 1.20 Using evidence markers to identify pedestrian contact locations.

© SAE International.

Figure 1.21 Pedestrian contact locations (close-ups).

(a)

(b)

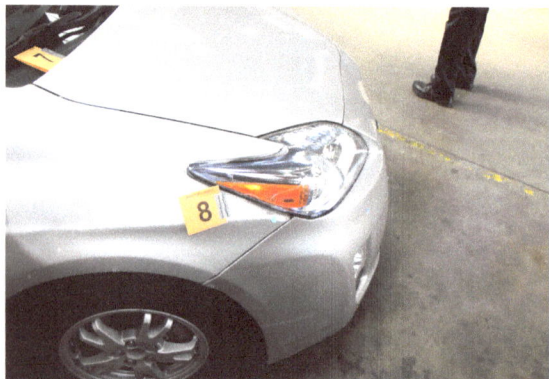
(c)

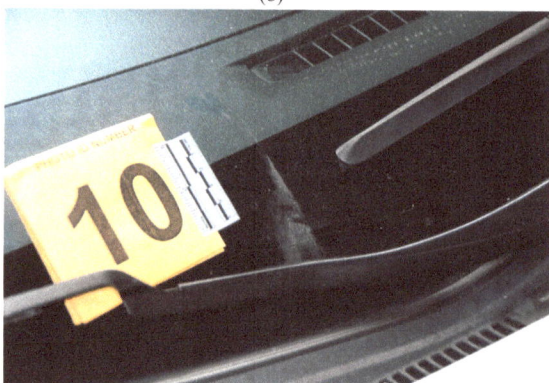
(d)

Figure 1.22 Video frames showing dummy contact points on Chevrolet Malibu.

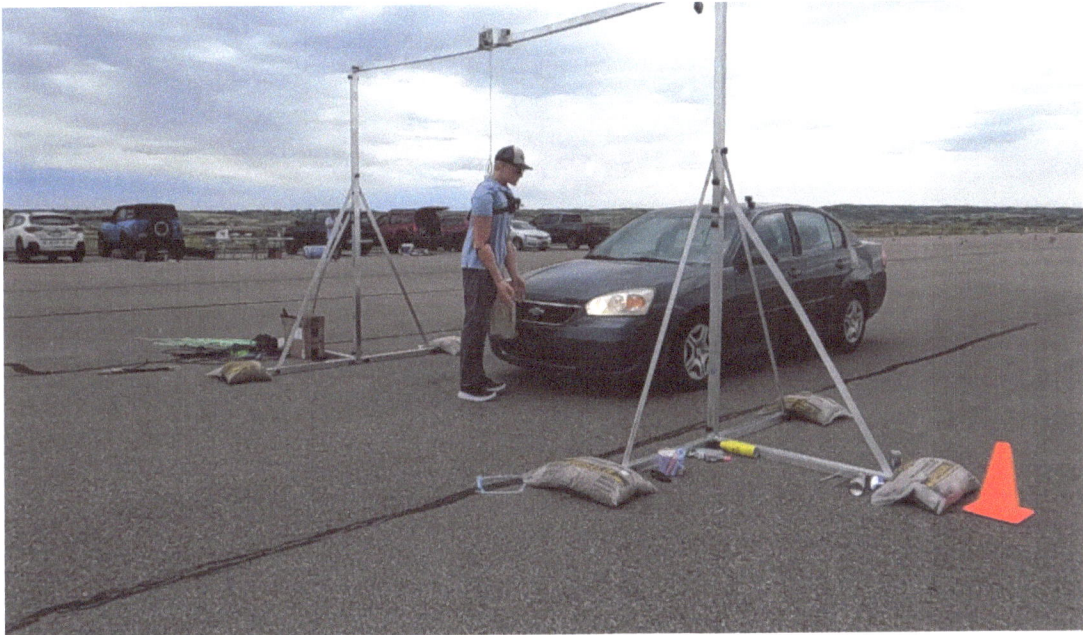

(a)

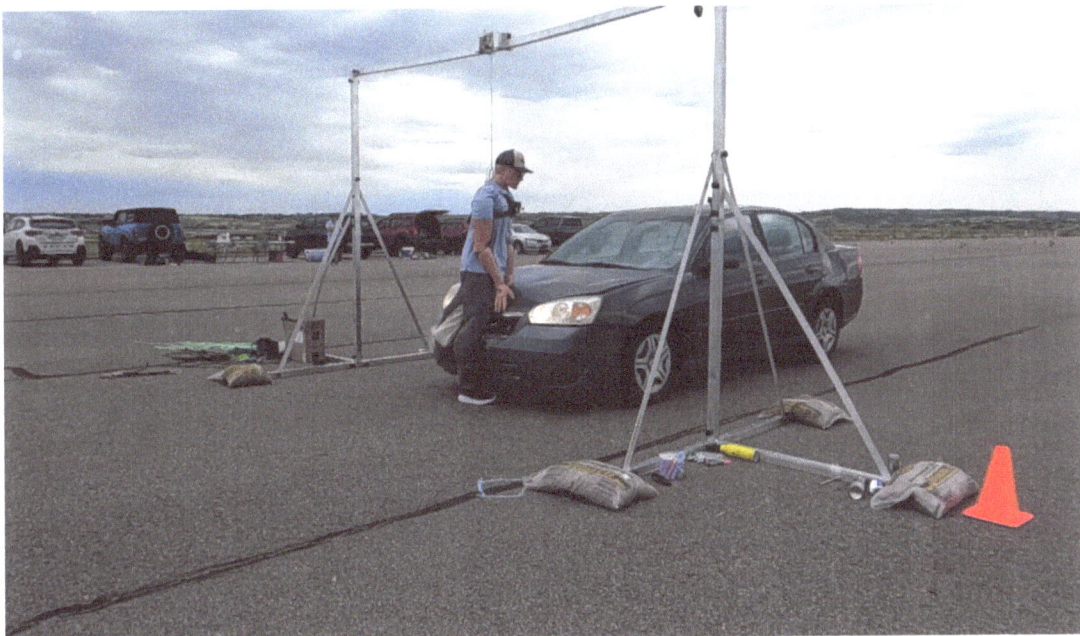

(b)

(Continued)

Figure 1.22 (Continued) Video frames showing dummy contact points on Chevrolet Malibu.

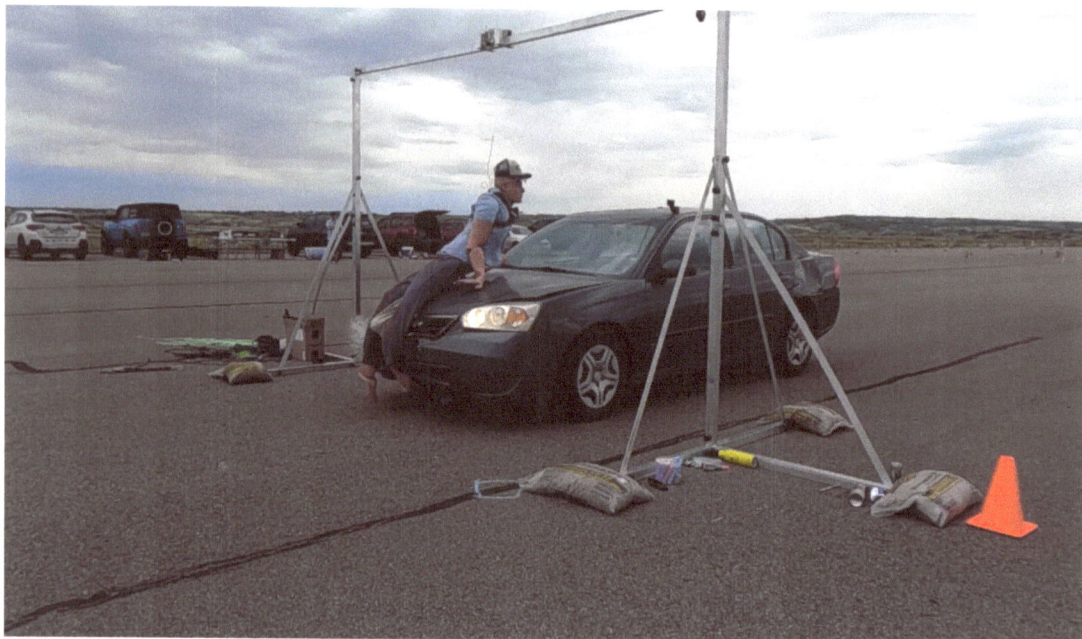

(c)

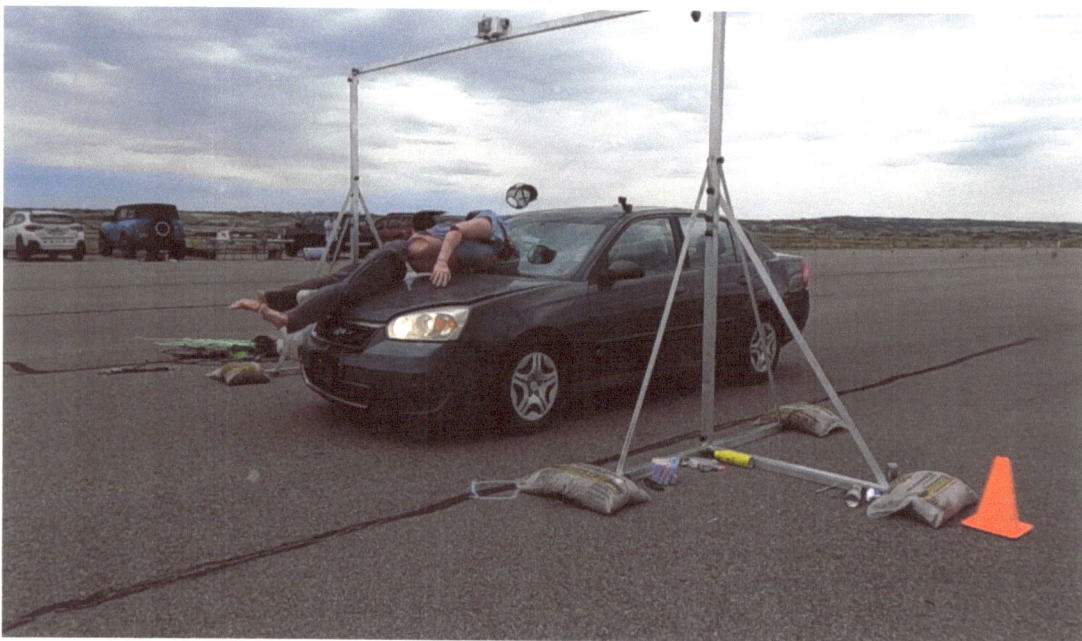

(d)

(Continued)

Figure 1.22 (Continued) Video frames showing dummy contact points on Chevrolet Malibu.

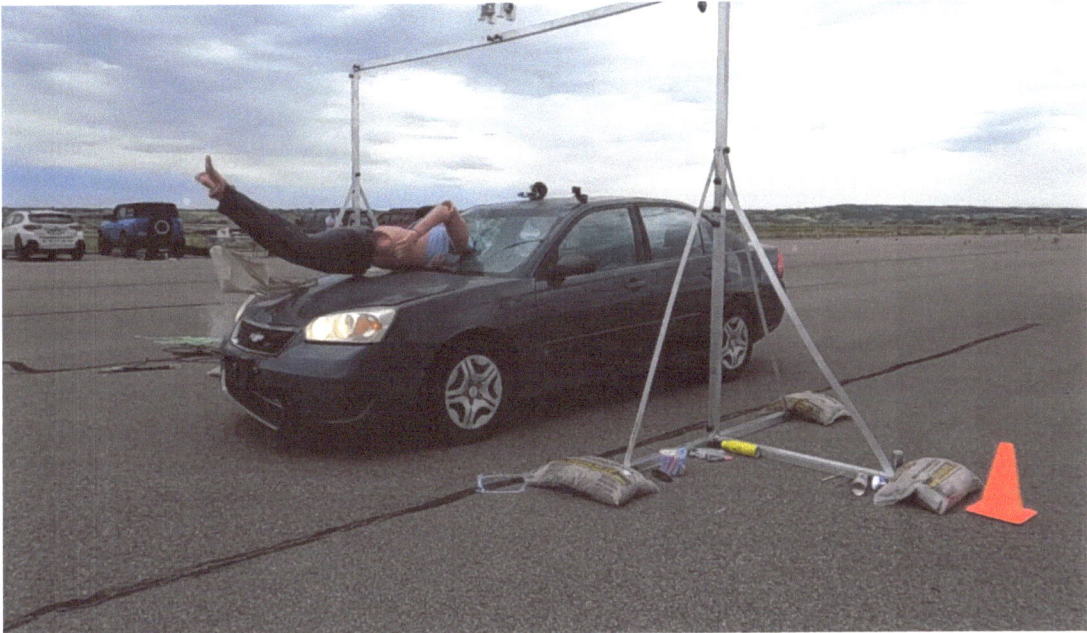

(e)

(f)

Table 1.1 Vehicle damage summary for forward projection trajectories (Happer, Table 1).

Approx. vehicle impact speed (km/h)	General damage summary
<20	Surface cleaning marks.
35	Leading edge of hood dented; deformation on front of vehicle.
60	Middle of hood dented.

Reprinted from Reference [1.12]. © SAE International.

Beyond assisting with identifying trajectory classification, the contact locations on the striking vehicle could potentially assist with speed determination. Happer et al. [1.12] reviewed vehicle damage information from the technical literature "to ascertain trends in the damage for specific vehicle impact speeds and collision circumstances. A set of guidelines were derived for estimating approximate vehicle impact speed from the sustained vehicle damage for both forward projection and wrap trajectories." These authors presented **Table 1.1** for forward projections and **Table 1.2** for wrap trajectories. These tables summarize the observed trends. For wrap trajectories, these authors observed that "the main indicator of vehicle impact speed is the location of secondary head contact…the adult pedestrian's head will typically make contact with the vehicle somewhere between the hood and the upper frame of the windshield." The higher the impact speed, the further back on the vehicle the pedestrian's head will strike. On the other hand, Happer et al. observed that "for a specific impact speed and pedestrian size, vehicles with different frontal profiles will likely induce different damage patterns. In addition, vehicle damage will differ for similar sized and shaped vehicles with unlike structural characteristics. Thus, vehicle structural differences and other collision factors (e.g., pedestrian size, pedestrian pre-impact motion, etc.) preclude exact assessment of vehicle impact speed from the generalized post impact vehicle damage. However, a damage analysis does allow an order of magnitude estimate of the vehicle impact speed."

Table 1.2 Vehicle damage summary for wrap trajectories (Happer, Table 2).

Approx. vehicle impact speed (km/h)	General damage summary
<20	Surface cleaning marks.
25	Head contact near bottom edge of windshield when pedestrian C.G. ~60 cm above low-fronted vehicle's bumper assembly; otherwise, head contact near middle of hood for average-sized vehicle and pedestrian.
	Body contact on roof when pedestrian C.G. ~85 cm above low-fronted vehicle's bumper assembly.
25–40	Head contact near trailing portion of hood or cowl; slight body panel deformation.
40	Head contact near bottom edge of windshield for impacts significantly below (~50 cm) pedestrian's C.G. (i.e., typical braking low-fronted vehicle).
40–50	Clearly defined dents on body panels.
50	Head contact near bottom edge of windshield when pedestrian C.G. ~40 cm above low-fronted vehicle's bumper assembly.
	Body contact on roof when pedestrian C.G. ~60 cm above low-fronted vehicle's bumper assembly.
50–55	Head contact near middle of windshield for typical braking low-fronted vehicle.
60	Head contact near bottom edge of windshield when vehicle's upper leading edge near pedestrian's C.G.
>60	More probable body to roof contact.
70	Head contact near upper frame of windshield; significant deformation of body panels.
80	Pelvic contact with roof; roof deformation (unbraked vehicle).

Reprinted from Reference [1.12]. © SAE International

Other authors have considered this method for obtaining an order of magnitude estimate of the striking vehicle's speed. According to Eubanks [1.13], "It is very difficult to accurately and precisely estimate the vehicle's speed based solely on the extent and nature of the damage to the vehicle. However, it is possible to correlate ranges of speed with types of marks or damage on the vehicle." Eubanks presented a series of trends similar to what Happer et al. presented. Eubanks continued: "In principle, the locations and angles of contacts should allow an impact speed to be determined. In practice, estimation of impact speed based on head contact position is very difficult and only ranges should be attempted. This is because of the variations in pedestrian heights, vehicle leading edge heights and lengths of hoods, and the lack of a large data base of crash test results. For a wide variety of accident conditions, the location of head contacts tends to be uniformly distributed along the full length of the hood. Is there a relationship between the speed of a vehicle and the location of the windshield head strike of the pedestrian? The answer is yes, depending on type and shape of vehicle and the height of the pedestrian. Generally, the likelihood that the pedestrian will reach the windshield varies directly with the pedestrian's height and the vehicle's speed, but inversely with the length of the vehicle's hood. For example, if the pedestrian were the height of a Wilt Chamberlain, 86 inches, then the pedestrian would reach the windshield at lower speeds than would the averaged sized pedestrian. Conversely, if the impacting vehicle is a Mercury Montego with a hood length of 65 inches, it would take a higher speed to cause a given pedestrian to strike the lower portion of the windshield." Eubanks also noted that the head impact location will be influenced by the deceleration of the striking vehicle.

In their book, Reade and Becker [1.14] are critical of the analysis represented in **Tables 1.1** and **1.2**. They state that, "in a perfect world where all vehicles were the same size, with the same frontal profile, and all pedestrians were the same height, there may be some reliable method to estimate where the pedestrian's head will first strike on the vehicle at any given speed. However, since all vehicles are not the same and all pedestrians are of differing heights, this claim cannot be supported with any degree of certainty. In the past, some information and graphics have circulated to show a relationship between the first head contact and the vehicle's striking speed. Some have even relied upon this information as the sole method to estimate a vehicle's speed at impact. Estimating vehicle speed is not quite that simple. If we can say anything about this approach to estimate vehicle speed, it would be this method is not reliable and it should be discouraged and never used to estimate vehicle speed." Our contention in this present book is that, conceptually, the method proposed by Happer et al. is valid. However, so is the criticism offered by Reade and Becker and the cautions noted by Eubanks. What is needed to make this approach viable is a method that can consider the particular vehicle shape involved in the collision when considering the influence of the vehicle speed on the pedestrian head impact location. Chapter 3 discusses a multibody simulation in depth, which is a method that can bridge the gap and provide the ability to consider the actual shape of the involved vehicle.

Post-Collision Decelerations for a Pedestrian

Theoretical models of pedestrian collisions require an input for the deceleration of the pedestrian following separation from the striking vehicle (the drag factor or coefficient of friction). Selecting a reasonable value for this input requires understanding how this parameter is defined within each model. The motion of a struck pedestrian can be segmented into four phases: (1) the impact phase; (2) the airborne phase; (3) the landing phase (impact with the ground); (4) and the sliding and tumbling phase. The total throw distance is the sum of the distance consumed by these phases. Sometimes, the drag factor is defined such that it includes only the sliding and tumbling phase. For the purposes of the discussion here, this will be referred to as Definition 1. In other instances, the drag factor is defined in a way that includes both the initial landing on the ground and the sliding and tumbling phase. This will be referred to as Definition 2. And finally, the drag factor is sometimes defined in a way that includes all four phases—the average deceleration of the pedestrian from first contact to rest. This will be referred to as Definition 3.

To explore the differences in these definitions, consider the statement by Severy and Brink [1.5] that a sliding or tumbling pedestrian "has a higher effective drag coefficient than the automobile undergoing emergency braking." This statement has been rendered generally untrue by improvements in the emergency braking capabilities of passenger vehicles. Even in 1966, this statement would have only been true if the speed loss from the pedestrian's impact with the ground after the airborne phase was included in the deceleration calculation (Definition 2). When the speed loss from the landing is included (but the airborne phase is excluded), high drag factors result. Schneider and Beier stated that the slide deceleration of a clothed pedestrian was approximately 0.8–1g [1.15]. Collins [1.8] stated that "the friction

coefficient for a sliding pedestrian who has been knocked down by an automobile is about 1.1. Because of the large impact forces and correspondingly high retarding forces to which a body is subjected when it first hits the ground, this number is higher than would be obtained if the pedestrian had not been knocked to the pavement. A rider launched from a motorcycle may fall to the ground or be slammed into the pavement, so on paved surfaces the ejected rider's friction coefficient will range between 0.8 and 1.2." As is evident from Collins' statements, these drag factors are high because they include the speed loss from the landing but exclude the airborne phase (Definition 2). Similarly, Haight and Eubanks [1.16] reported decelerations between 0.8 and 0.95g for anthropomorphic test devices (ATDs) sliding or tumbling on the ground after being involved in vehicle versus bicycle collisions. Hill [1.17] reported experiments in which "a life size dummy was constructed using a leather motorcyclist's suit stuffed with a mixture of sand and sawdust contained in bags." The dummy was dressed in various types of clothing, and a minimum of ten tests were completed with each set. "In each test the dummy was held, face up, horizontal to and approximately 10 cm above the ground…On release of the dummy the speed at release and the first point of contact with the road is known, the slide and/or roll displacement could be measured and the effective coefficient calculated." Hill calculated an average drag factor of 0.8 for his tests.

On the other hand, Fricke [1.6, 1.7] reported the following drag factors for pedestrians sliding on various surfaces: grass, 0.45 to 0.70; asphalt, 0.45 to 0.60; concrete, 0.40 to 0.65. These drag factors are significantly lower than those in the previous paragraph because they exclude the speed loss the pedestrian experiences from landing after the airborne phase. In fact, these values were obtained by simply dragging a person along the ground, without any airborne or landing phases. Han and Brach [1.18] adjusted Hill's calculations and reported an adjusted drag factor (removing the effects of the ground impact) of approximately 0.76. Wood and Simms [1.19] had previously corrected Hill's friction coefficients to remove the effects of the initial ground impact and reported a 95% confidence range for the mean value of 0.679 to 0.759. The Han and Brach calculations may have insufficiently adjusted Hill's values. Regardless, the point remains that when the impact with the ground is excluded, lower drag factors result. Wood and Simms cited additional studies as well, all of which reported lower values for the drag factor than those reported by Hill. For example, Searle's [1.20] data exhibited a mean value of 0.504, and the Bovington dataset reported by Craig [1.21] had a mean value of 0.584.

Wood and Simms ultimately concluded that "the value of μ from experimental tests is subject to scatter. The range of μ from the Searle, Hill, and Bovington test series is $\mu = 0.39 - 0.87$. By comparison, the 99% confidence range from previously reported work is $\mu = 0.3$ to 0.93." Both ranges have a middle value just over 0.6. Happer [1.12] conducted a literature review and explored all three definitions for the drag factor. He reported a range of drag factors for Definition 1 of 0.40 to 0.72 (a narrower range, but consistent with what Wood and Simms reported), for Definition 2 of 0.7 to 1.22, and for Definition 3 of 0.37 to 0.79. Haight and Eubanks [1.16] had reported a range of drag factors of 0.24 to 0.47 when the total throw distance was used in analyzing the motion of the dummies in their bicycle collisions (Definition 3). It is interesting that the ranges for Definitions 1 and 3 are overlapping and similar. The reason for this can

be discerned by considering this lengthy quotation from Searle [1.20]:

A pedestrian, after being projected by the impact, will normally undergo a trajectory through the air, then a period of ground contact, then possibly further trajectories through the air interspersed with periods of ground contact. A useful parallel with this process, and helpful in understanding it, is the skip skids sometimes seen when a locked wheel bounces off the road whilst skidding. That can happen due to a bump on the road, giving a single skip in the skid, or due to the repeated bounding of a trailer wheel.

The standard advice given when measuring skip skids is to measure end to end as if there were no breaks in the skid. A moment's reflection shows why that is the correct basis for calculating speed. During the whole skid, end to end, the average reaction from the ground must have been equal to the weight on the wheel. Were that not so, the net vertical force on the wheel would have given it a vertical velocity and a vertical displacement by the end of the skid. The reaction from the ground has varied, but any reduction whilst the wheel is out of contact must have been made up by an increase whilst it is in contact. The same must apply to the frictional force, which at any instant is simply μ times the ground reaction. The average frictional force will therefore be μ times the average ground reaction, or in other words μ times the weight on the wheel, exactly the same as if there had been a continuous skid.

The same principles apply when considering the average frictional force acting on the projected object.

During periods of ground contact, the reaction of the ground is high to make up for the periods spent in the air. The reaction is particularly high when the pedestrian first lands after a prolonged time out of contact.

These paragraphs from Searle have several implications. First, the absence of speed loss during the initial airborne phase is offset by the presence of high speed loss from the landing. Second, the deceleration that occurs for pure sliding of a pedestrian is similar to what would occur for a pedestrian who is bouncing or tumbling with alternating periods of airborne motion and ground contact. This particular point has broader applicability in accident recon-struction [1.22]. Imagine, for example, a motorcycle that is intermittently marking the roadway as it slides and tumbles. In other words, over the distance from when the motorcycle begins to capsize to the point where it comes to rest, there is variation in the degree to which the motorcycle is engaging with the surface, and perhaps there are even periods of airborne motion. In this case, the average deceleration should be applied over the entire distance, and the value of deceleration is not diminished by periods of lesser engagement with the ground because these periods of lesser engagement are offset by periods of more intense engagement. Third, from the point when the pedestrian separates from the striking vehicle (not neces-sarily the same as the point at which first contact occurs) to the point where the pedestrian comes to rest, the average deceleration is about the same as the deceleration for just the sliding and tumbling portion of the motion.

One caveat: while the deceleration for tumbling and bouncing is about the same as the

deceleration for pure sliding, pure rolling is a different matter. This is generally not a concern when analyzing pedestrian crashes, but with other crash types, the distinction could become important. During a high-speed rollover crash, for example, the average deceleration over the entire roll phase is likely to be about 0.44g (plus or minus) [1.23]. However, if we look at the instantaneous deceleration, the deceleration in the early portions of the roll phase is higher than in the later portions [1.24]. The reason is largely the transition from sliding, bouncing, and tumbling to pure rolling [1.25]. Similarly, for a lower-speed rollover, where the roll phase might consist entirely of pure rolling without sliding, the average deceleration is likely to be lower— say 0.25g (plus or minus) [1.26].

Based on his literature review, Happer also observed that the drag factor for a sliding or tumbling pedestrian is independent of the surface type and that it does not change if the surface is wet. Wood [1.27] also reported no evident difference in friction between dry and wet conditions. Fugger [1.28], on the other hand, reported a series of 140 pedestrian crash tests utilizing high-fronted vans that would generate forward projection trajectories. The dummy, which was 75.5 in. tall and weighed 169 lb, was clothed in a wetsuit covered by coveralls and standard athletic footwear. Of the 140 tests, 56 were conducted on dry asphalt and 84 on wet asphalt. Impact speeds varied between 4 and 60 km/h, with most of the tests conducted at speeds below 32 km/h. Fugger found that the dummy throw distances were generally longer on a wet road. He reported drag factors for the wet surfaces between 0.31 and 0.41.

Sullenberger [1.29] conducted testing of pedes-trian dummies sliding and tumbling on low-friction surfaces, such as snow, ice, or plastic sheeting. The dummies were either impacted by an automobile, launched from a ramp at angles between 10 and 20°, or drug across the testing surface. Drag testing on plastic sheeting produced a drag factor of approximately 0.36. From the ramp testing, Sullenberger stated that "the measured drag factor of the packed snow and the black plastic sheeting ranged from 0.39 to 0.49." Later in this article, the author appears to state that the drag factor for the snow was 0.39 and that the drag factor for the plastic sheeting was 0.49. Drag testing with a human subject clothed in insulated cotton overalls yielded the following drag factors: rough ice = 0.213 to 0.255; smooth ice = 0.277 to 0.298; loose to packed snow = 0.378 to 0.459; 6-in.-deep loose snow = 0.513 to 0.621; and slush on asphalt = 0.324 to 0.351.

In addition to the potential influence of surface conditions, Hill's data showed that the drag factor can be influenced by clothing type. Hill noted: "The first twenty tests were conducted with the dummy wearing jacket and trousers made of serge material (Police Uniform). The range of calculated friction coefficient was 0.699–0.926. In two of the following series of tests, there was no appreciable change in the range of calculated coefficient of friction. The clothing consisted of terelyne jumper, trousers and body warmer, and then a woolen boiler suit. The calculated range for this series was 0.707–1.065…A set of ten tests were conducted with the dummy wearing a nylon motorcyclist suit, these tests produced a lower range, 0.573–0.716." Han and Brach analyzed Hill's data and found that the differences between clothing types were statistically significant, though they concluded that "only nylon showed a meaningful change."

Theoretical Models of Pedestrian Collisions

The technical literature contains a large number of theoretical models for the post-collision motion of a pedestrian. If the takeoff (immediately after separation from the vehicle) and landing locations of the pedestrian were known, then it would only be necessary to model the airborne portion of this motion. Typically, though, neither of these is known, and so most theoretical models incorporate the airborne, landing, tumbling, and sliding phases. Toor [1.30] has noted that these theoretical models "are very difficult to apply to real world collisions, because the data necessary to solve the mathematical equations is only partly available from the real world collisions." Similarly, Fricke [1.6] observed that "investigators often try to estimate the velocity of a car by applying the vault equation to the pedestrian. Usually this will not yield acceptable answers, because too many simplifying assumptions are used." Still, theoretical models can help reconstructionists to identify which parameters are significant to a reconstruction, and they can give them physical insight into how these crashes unfold. As Searle [1.20] has noted, "field studies by themselves provide no understanding of the physical process. That understanding is important in the consideration of the effects of site characteristics, such as gradient or coefficient of friction." This section covers several theoretical models that have appeared in the collision reconstruction literature and highlights physical principles and trends that can be drawn out with these models. The intent of this section is not to be exhaustive, but rather to give the reconstructionist a sense of the main components of these models, how they differ, and what insights can be gained from them.

Collins Model

Collins [1.8] proposed Equation (1.1) for calculating the launch speed of a struck pedestrian who first undergoes an airborne trajectory and then lands and slides to rest. In this equation, V is the pedestrian throw speed in miles per hour, H is the height in feet of the pedestrian's CG above the ground immediately prior to the collision, μ is the coefficient of friction between the pedestrian and the road surface, and S_T is the pedestrian throw distance in feet. This model yields the projection speed of the pedestrian, not the impact speed of the vehicle, and air resistance is neglected. This equation also assumes zero vertical velocity at the beginning of the airborne phase (no takeoff angle).

$$S_T = \frac{V\sqrt{H}}{2.73} + \frac{V^2}{30\mu} \tag{1.1}$$

Searle [1.20] was critical of Collins' assumption that the struck pedestrian would not have any vertical velocity after separation from the vehicle. He noted: "One only has to look at film of pedestrian impact tests to know that such an assumption is not justified: the pedestrian is usually lofted by the inclined front of the car… [The assumption of no initial vertical velocity] is demonstrably incorrect in most cases, so that the Collins formula gives no real appreciation of the physics involved." Searle also noted that the Collins model ignored the speed loss from the pedestrian landing after the airborne phase. In Searle's discussion of the Collins model, Searle gives a more general form of Equation (1.1), as follows:

$$S_T = V\sqrt{\frac{2H}{g}} + \frac{V^2}{2\mu g} \tag{1.2}$$

In this equation, g is the gravitational constant in units of feet per second squared. In this form of the equation, velocity would be entered in feet per second rather than in miles per hour.

Searle also notes that, in his discussion of projected motorcycle riders, Collins made allowance for an upward projection velocity of the rider, with the following equation:

$$S_T = V\sqrt{\frac{2\Delta H}{g}} + V\sqrt{\frac{2H}{g}} + \frac{V^2}{2\mu g} \tag{1.3}$$

In this equation, ΔH is the difference between the height of the pedestrian's CG above the ground at the first contact and at the apogee of the post-impact trajectory. Thus, Equation (1.3) requires the reconstructionist to know the apogee height. Searle observes that "unfortunately the apogee height is seldom recorded, other than by unreliable guesses."

Searle Model, 1983

Searle [1.31] derived Equation (1.4) for determining a pedestrian's projection velocity, based on the conceptual model depicted in **Figure 1.23**. In this equation, μ is the coefficient of friction between the pedestrian and the ground, g is the acceleration due to gravity, s is the throw distance, including the airborne and sliding and tumbling phases, and θ is the projection angle.

Searle reported a typical coefficient of friction for a pedestrian sliding on asphalt of 0.66. He claimed that this coefficient of friction would not be different if the surface was wet, but this has not consistently been borne out by later experimental data (see the discussion above). When comparing his formulas to experimental data, Searle applied a range on the coefficient of friction between 0.3 and 1.0. In his later study discussed [1.20], Searle reported a typical coefficient of friction of 0.7. The Searle model improved on the Collins model by incorporating a launch angle and by incorporating the ground plane speed loss due to the pedestrian landing on the ground after vaulting through the air.

Figure 1.23 Searle model.

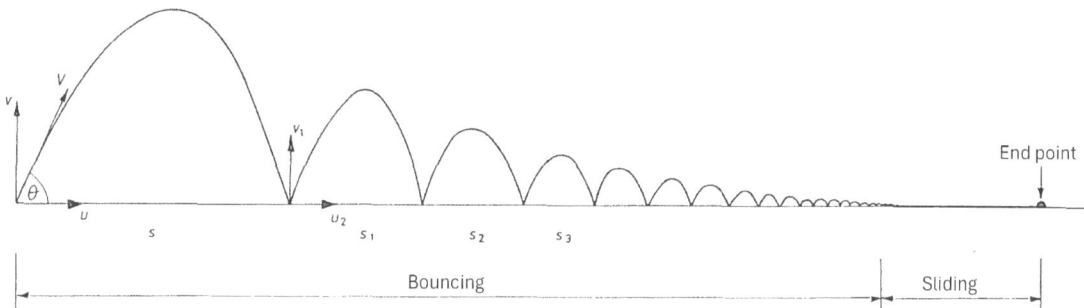

$$v_{proj} = \frac{\sqrt{2\mu gs}}{\cos\theta + \mu\sin\theta}$$ (1.4)

This formula does not yield the impact speed of the vehicle, but rather the speed at which the pedestrian is accelerated as a result of the collision. Often, the pedestrian will not be accelerated up to the full speed of the car. To determine the impact speed of the vehicle, it is helpful to introduce the concept of projection efficiency, which is defined as follows. In this equation, v_{proj} is the projection speed of the pedestrian, v_{impact} is the impact speed of the vehicle, and η_{proj} is the projection efficiency.

$$\eta_{proj} = \frac{v_{proj}}{v_{impact}}$$ (1.5)

Incorporating the projection efficiency, Equation 1.6 would yield the speed of the vehicle. Searle reported a typical projection efficiency of 0.640 for adults impacted by a low-fronted vehicle, 0.744 for adults impacted by a high-fronted vehicle, 0.727 for children impacted by a low-fronted vehicle, and 0.831 for children impacted by a high-fronted vehicle. Happer reported projection efficiencies for forward projections between 0.8 and 1.0 and for wrap trajectories between 0.5 and 0.92. In practice, these large ranges of projection efficiency are an unknown that introduces significant uncertainty into a calculation using the theoretical models described in this section.

$$v_{impact} = \frac{1}{\eta_{proj}} \frac{\sqrt{2\mu gs}}{\cos\theta + \mu\sin\theta} \tag{1.6}$$

For their forward projection tests on the dry surface, Fugger et al. [1.28] reported an average takeoff angle for the pedestrian of 1.78°, with a standard deviation of 2.1°. For the tests on the wet surface, they reported an average takeoff angle for the pedestrian of 2.6°, with a standard deviation of 2.7°. When reconstructing real-world collisions, the projection angle is often not known, and hence, Searle developed equations representing the minimum and maximum values that Equations (1.6) could yield. These equations are as follows:

$$v_{impact,min} = \frac{1}{\eta_{proj}} \sqrt{\frac{2\mu gs}{1+\mu^2}} \tag{1.7}$$

$$v_{impact,max} = \frac{1}{\eta_{proj}} \sqrt{2\mu gs} \tag{1.8}$$

Application of Equation (1.8) requires the analyst to ensure that the maximum projection angle could not have been greater than a critical projection angle calculated using the following equation. This will typically be the case for pedestrian collisions.

$$\theta_{crit} = 180° - 2\arctan\frac{1}{\mu} \tag{1.9}$$

For a coefficient of friction of 0.4, the critical projection angle is 44°, and for a coefficient of friction of 0.7, the critical projection angle is 70°. For a vehicle collision with a pedestrian, it is unlikely that these projection angles would be exceeded.

Aronberg Model

Aronberg [1.32] derived another model. His model differed from the Collins model by utilizing "the time of the body in the air, at the launch velocity, to determine the adjustment necessary to the launch velocity to account for air drag. In addition, the component of velocity corresponding to total horizontal distance traveled is limited to the horizontal velocity component of the body following the airborne trajectory." Aronberg noted that "Collins' method incorrectly includes both the vertical and horizontal components." The Aronberg model is shown graphically in **Figure 1.24**, along with the meaning of the variables within the model.

Figure 1.24 Aronberg model.

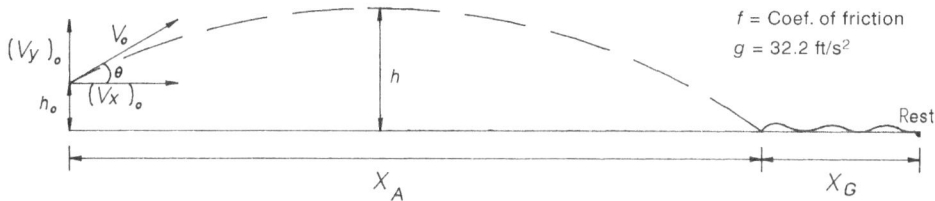

From this model, Aronberg derived and presented the following equation relating the total throw distance, d_t, to the projection speed:

$$d_t = \frac{V_0^2 \cos\theta \sin\theta + \sqrt{\left(V_0 \sin\theta\right)^2 + 4\frac{g}{2}h_0}}{g} + \frac{V_0^2 \cos^2\theta}{2gf} \qquad (1.10)$$

where

f is the coefficient of friction

g is the gravitational constant

V_0 is the projection velocity

θ is the projection angle

Like the Collins and Searle models, the Aronberg model yields the projection velocity of the pedestrian, not the impact speed of the striking vehicle. The projection efficiency would need to be incorporated in order to arrive at the vehicle impact speed. The implementation of the projection efficiency would be carried out identically to the way it was implemented in the Searle model. Aronberg's discussion of air resistance will be covered later in this section.

Searle Model, 1993

Searle [1.20] expanded Equation (1.4) to account for the difference in the CG height of the pedestrian at the end of vehicle contact and the height at rest. This difference is given as H in Equation (1.11) and is typically a negative value to account for the decrease in a pedestrian's CG height from when they are initially standing upright and then laying down at rest. The height of the pedestrian at the end of contact will usually be estimated as the standing CG height of the pedestrian just prior to impact, though a different value could be used if the reconstructionist has information or evidence leading to a different assumption.

$$v_{proj} = \frac{\sqrt{2\mu g\left(s + \mu H\right)}}{\cos\theta + \mu\sin\theta} \qquad (1.11)$$

Searle also expanded the minimum projection velocity to account for this height distance by differentiating with respect to θ and equating to zero. The resulting angle is equal to $\tan^{-1}\mu$. Substituting this value into Equation (1.11) yields Equation (1.12). Keep in mind that the projection efficiency would need to be incorporated into Equations (1.11) and (1.12) in order for them to yield the vehicle impact speed. Without that, they yield the pedestrian's projection velocity.

$$v_{proj,min} = \sqrt{\frac{2\mu g(s + \mu H)}{1 + \mu^2}}$$

(1.12)

To test this model, Searle performed laboratory tests and pedestrian dummy drop tests. The laboratory tests involved sliding a small flat object approximating and idealized particle down a pair of polished chrome rails and measuring the travel distance of the object as it slid across surfaces with various friction coefficients. For dummy tests, Searle dropped a dummy from the top of an open double-deck bus at various horizontal speeds up to 35 mph and measured the travel distance after landing. **Figure 1.25** shows the test setup used by Searle to perform the drop tests. Searle reported that his tests produced results that were in good agreement with Equation (1.11).

Figure 1.25 Pedestrian dummy drop test by Searle.

One caveat within the results: within the Searle model (and other theoretical models), the assumption is made that the landing would produce a change in horizontal velocity equal to the following:

$$\Delta u = \mu v$$

(1.13)

where

Δu is the ground plane (horizontal) change in velocity

v is the vertical velocity at landing

μ is the coefficient of friction

For the set of dummy drop tests conducted by Searle, he reported that the actual value of the change in horizontal velocity was approximately 63% of what this equation predicted. This reportedly did not significantly degrade the accuracy of Equation (1.11).

Han–Brach Model

Han and Brach [1.18] presented a theoretical model to calculate the vehicle impact speed based on the pedestrian throw distance. Similar to other models, the Han–Brach model included phases for the airborne, landing, and sliding and tumbling phases. In addition to these phases, these authors introduced an additional phase at the beginning that included the distances the vehicle and the pedestrian travel during the timeframe from the first contact between the vehicle and the pedestrian to secondary contact and separation. For a wrap trajectory, this would be the distance the vehicle travels from the time it first touches the pedestrian's leg to the time the pedestrian wraps onto and impacts the hood, and perhaps the windshield, of the vehicle. The definitions of the variables used in this model are illustrated in **Figure 1.26**.

Figure 1.26 Variable definitions for the Han–Brach model.

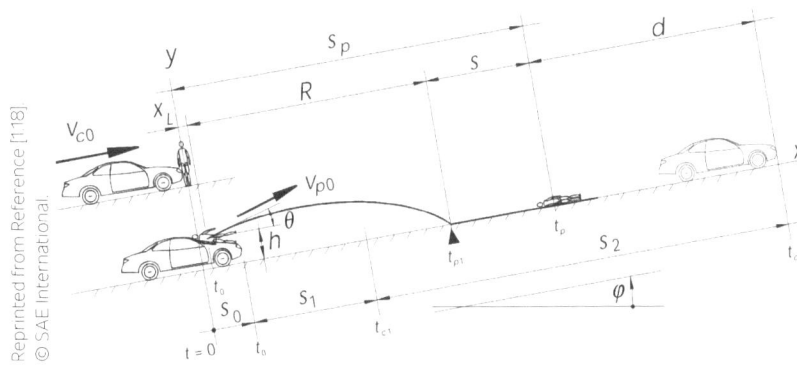

In the Han–Brach model, x_L is the distance traveled by the pedestrian while in contact with the vehicle, R is the distance traveled by the pedestrian in the air, and s is the distance the pedestrian slides/tumbles along the ground. The sum of these three distances is the total throw distance, s_p.

$$s_p = x_L + R + s \tag{1.14}$$

Han and Brach presented the following equations for calculating the initial speed of the car, v_{c0}:

$$v_{c0} = A\sqrt{s_p - B} \tag{1.15}$$

In Equation (1.15), the variables A and B are defined as follows:

$$A = \frac{m_c + m_p}{\alpha m_c} \sqrt{\frac{2 f_p g}{f_p^2 \sin^2 \theta + f_p \sin 2\theta + \cos^2 \theta}} \tag{1.16}$$

$$B = x_L + f_p h \tag{1.17}$$

where

m_c is the mass of the car

m_p is the mass of the pedestrian

α is the projection efficiency

f_p is the drag factor for the pedestrian

θ is the launch angle

g is the gravitational constant

h is the CG height of the pedestrian above the ground at the moment the pedestrian separates from the vehicle and is launched

The primary innovation of the Han–Brach model is the inclusion of an additional phase—the distance covered by the vehicle and pedestrian during the period of contact. Given how infrequently theoretical models are used in practice, this could be seen as an unnecessary complication to the modeling. Alternatively, given how infrequently these models are used, we might as well go hog-wild and include as many factors and phases as we can think to include and see what physical insight can be gleaned. When these authors fit their model to experimental data, several conclusions emerged, some of which are consistent with prior literature and some which are not. Han and Brach cited the data published by Hill [1.17] in which the average drag factor was reported as 0.8. Han and Brach corrected the Hill data to exclude the speed loss of the dummy from landing on the road surface and arrived at an adjusted average drag factor of approximately 0.76. Wood and Simms [1.19] had previously corrected Hill's drag factors to exclude the initial impact with the ground and reported a 95% confidence range for the mean value of 0.679 to 0.759. The Han–Brach model may have insufficiently adjusted Hill's values. Still, even corrected Hill's values are on the higher end of the drag factors other studies have reported for sliding and tumbling pedestrians. Han and Brach recognized this, noting that "specific reasons for these differences are not known." Despite recognizing that Hill's values were on the higher end, the Hill dataset was the only one referenced by Han and Brach in establishing inputs for their calculations and modeling.

Air Resistance

There will be some speed loss from air resistance during the airborne phase. However, this speed loss is likely small, and air drag is often neglected in theoretical models of a pedestrian's post-impact motion. The empirical models discussed in the next section inherently account for air resistance. Collins stated that speed loss from air resistance could be reasonably neglected at projection speeds below 25 mph (40 km/h). At projection speeds exceeding 25 mph, he stated that the calculated speed from his model should be corrected to account for air resistance by adding a correction to the calculated speed. This approach is essentially never used in practice.

The Aronberg model also neglected speed loss due to air resistance. In order to incorporate this speed loss, he examined published data on skydiving free fall velocity with respect to time. Aronberg noted that that the speed loss due to air resistance generally increases linearly up to a speed around 60 mph. At a speed of 61 mph, air resistance would cause just under 3 mph of speed loss for every second the pedestrian is in the air. Above 61 mph, the air resistance continues to increase, but at a greater rate than at speeds below 61 mph. Aronberg notes that "the free fall data presented and air drag derived are for humans in a spread position" and, therefore, likely represents the maximum expected air drag. Aronberg observed that his data yielded corrections significantly lower than what Collins suggested. The Collins method overestimates the speed loss due to air resistance.

Empirical Models for Determining Vehicle Speed from Throw Distance

The complexity of these theoretical models will be gratuitous for many reconstructions of pedestrian collisions. Empirical models are simpler and require a single input—the throw distance of the pedestrian. The use of throw distance to estimate the striking vehicle speed goes back decades in the literature. Schneider and Beier [1.15] examined both the longitudinal (along the travel direction of the vehicle) and lateral (transverse to the travel direction) throw distance of struck pedestrians. They observed that a relationship between throw distance and speed only existed when the striking vehicle was being braked at impact. If the vehicle was not being braked, the pedestrian could get carried or run over, and throw distance would become

unpredictable. Stcherbatcheff [1.10] similarly noted that "the distance between the point of collision…and the point at which the dummy comes to rest is essentially linked to two parameters, namely: the collision speed [and] the intensity of the braking of the vehicle…violent braking cuts down the contact time between the vehicle and the dummy, and contact with the ground occurs earlier, with the dummy in a more unfavorable attitude and moving with a higher horizontal speed component. The point at which the dummy comes to rest also lies further from the front end of the vehicle than in the case of moderate braking."

Schneider and Beier also distinguished between central collisions and out-of-center collisions. If the pedestrian was struck by the front end of the vehicle within 12 in. of the edge of the vehicle, then the collision was classified as an out-of-center collision. Otherwise, it was classified as a central collision. For central collisions, Schneider and Beier found that there was no significant influence of the pedestrian's weight on the resulting throw distance. They also found that "no remarkable effect of front-end design can be recognized with respect to the throw distance." Finally, they reported that the longitudinal throw distance was not influenced by the travel direction or speed of the pedestrian, but the lateral throw distance "is decisively influenced by the pedestrian's direction and speed… the pedestrian who was hit within the middle of the vehicle front continued to move in the original direction of walking."

For out-of-center collisions, Schneider and Beier noted that "a distinction had to be made as to whether the pedestrian was moving towards or moving away from the center of the vehicle front end…one remarkable result is that the [lateral] throw distance depends substantially on whether the pedestrian is hit by the automobile when he is moving toward or away from its center…the pedestrian who is moving toward the center is normally thrown in a direction [laterally] opposite to his original direction of movement or – depending on how far away he was from the car edge – he will scarcely be thrown in his original [lateral] direction of movement. If the pedestrian is hit moving away, his [lateral] throw direction will be identical to that in a central collision; he maintains his direction of movement."

In a study published in 2000, Happer [1.12] compiled and combined throw distance data from previously reported staged and real-world collisions and presented linear regression equations relating the pedestrian throw distance to the vehicle impact speed for both wrap and forward projection trajectories. For forward projection trajectories, Happer compiled a total of 106 collisions and found that the adult and child trajectory data could be combined. For wrap trajectories, Happer compiled a total of 202 collisions. Consistent with prior treatments, Happer found that "there is reduced correlation between impact speed and throw distance for non-braking vehicles. This variance is likely due to the pedestrians being carried on the vehicles for further distances before separation… consequently, the ten datapoints corresponding to unbraked vehicles were omitted for the remaining analysis." Happer reported the following regression equations for the data. In these equations, d_t is the throw distance in meters and v_v is the vehicle impact speed in kilometers per hour. The plus/minus reported for these equations is the 15th to 85th percentile range.

Forward Projection Trajectories

$$v_v = 11.4\sqrt{d_t} - 0.4 \quad \left[\pm 10.5\,\text{km/h}\right] \tag{1.18}$$

Wrap Trajectories

$$v_v = 12.7\sqrt{d_t} - 2.6 \quad \left[\pm 9.0\,\text{km/h}\right] \tag{1.19}$$

An English unit version of Equations (1.18) and (1.19) is included below. In these equations, d_t is the throw distance in feet and v_v is the vehicle impact speed in miles per hour.

Forward Projection Trajectories

$$v_v = 7.1\sqrt{0.305 \cdot d_t} - 0.25 \quad \left[\pm 6.5\,\text{mph}\right] \tag{1.20}$$

Wrap Trajectories

$$v_v = 7.9\sqrt{0.305 \cdot d_t} - 1.6 \quad \left[\pm 5.6\,\text{mph}\right] \tag{1.21}$$

From these regression equations, Happer concluded that "the average drag factor for a pedestrian in a forward projection case is about 0.5…The average drag factor for a pedestrian in a wrap trajectory was evaluated to be about 0.6."

In a study published in 2001, Randles et al. [1.33] studied real-world pedestrian collisions captured from a camera in a bus station clock tower "overlooking a busy downtown intersection" in Helsinki, Finland. The camera had been placed in this clock tower in February 1991, and at the time of the Randles et al. study, it had captured 15 vehicle–pedestrian collisions. The footage from 13 of these collisions "was analyzed using digitizing motion analysis software to quantify the pre-impact and post-impact trajectories of both the vehicle and the pedestrian for each accident." These 13 collisions included four fender vaults, seven wrap trajectories, and two forward projections. Randles compared the throw distances from these collision to other data and models in the literature. He concluded that "the throw distances from the real-life wrap trajectories correlated well with the pedestrian

crash testing." The exception was the two instances where the impacting vehicle was not decelerating during the impact and the pedestrians were carried.

In 2002, Toor et al. [1.34] first tested Happer's equations using data from 51 real-world crashes reported by Hill [1.15], Dettinger [1.35], and Randles et al. [1.33]. These real-world collisions were not part of Happer's original regression analysis. Toor et al. reported that the Happer model "accurately predicts vehicle impact speeds in real world pedestrian collisions." After demonstrating this, Toor et al. incorporated the real-world data into Happer's regression analysis and reported updated regression equations. This resulted in the following equation for forward projection collisions. In these equations, the throw distance is expressed in meters and the resulting speed in kilometers per hour.

$$v_v = 11.3\sqrt{d_t} - 0.3 \quad [\pm 10.5\,\mathrm{km/h}] \tag{1.22}$$

For wrap trajectory, Toor et al. reported the following equation:

$$v_v = 13.3\sqrt{d_t} - 4.6 \quad [\pm 9.0\,\mathrm{km/h}] \tag{1.23}$$

Following is an English unit version of these equations. In these equations, the throw distance is expressed in feet and the resulting speed in

miles per hour. The following equation is for forward projections:

$$v_v = 7.0\sqrt{0.305 \cdot d_t} - 0.2 \quad [\pm 6.5\,\mathrm{mph}] \tag{1.24}$$

The following equation is for wrap trajectories:

$$v_v = 8.3\sqrt{0.305 \cdot d_t} - 2.9 \quad [\pm 5.6\,\mathrm{mph}] \tag{1.25}$$

In 2002, Fugger et al. [1.27] reported a series of 140 pedestrian crash tests utilizing high-fronted

vehicles that would generate forward projection trajectories. An Alderson Research Labs CG-95 dummy was utilized (75.5 in. tall, 169 lb). The dummy was clothed in a wetsuit covered by coveralls and standard athletic footwear. The following vans were used in this test series: a 1976 Ford Econoline 250, a 1971 Dodge B200, a 1982 Dodge B250, a 1980 Plymouth D100, a 1982 Chevrolet G20, and a 1977 Dodge Sportman. The leading edge of the hood of each of these vehicles was above the CG of the dummy such that collisions would produce forward projection trajectories. Of the 140 tests, 56 were conducted on dry asphalt and 84 on wet asphalt. Impact speeds varied between 4 and 60 km/h, with most of the tests conducted at speeds below 32 km/h. Fugger et al. reported that the dummy throw distance was generally longer under wet roadway conditions. Fugger et al. reported the following equations for these forward projection collisions. In these equations, d_t is the throw distance in meters, CG is the height of the pedestrian's CG prior to the collision in meters, and S is the vehicle speed in kilometers per hour.

Dry Roadway, Low- and High-Speed Estimates

$$S = \binom{8.77}{13.76} \cdot \sqrt{d_t - CG} \tag{1.26}$$

Wet Roadway, Low- and High-Speed Estimates

$$S = \binom{8.77}{13.76} \cdot \binom{0.31}{0.41} \sqrt{d_t - CG} \cdot \left(\frac{1}{0.43}\right) \tag{1.27}$$

An English unit version of these equations is included below. In these equations, d_t is the throw distance in feet, CG is the height of the pedestrian's CG prior to the collision in feet, and S is the vehicle speed in miles per hour.

Dry Roadway, Low- and High-Speed Estimates

$$S = \binom{5.45}{8.54} \cdot \sqrt{0.305 \cdot (d_t - CG)} \tag{1.28}$$

Wet Roadway, Low- and High-Speed Estimates

$$S = \binom{5.45}{8.54} \cdot \binom{0.31}{0.41} \sqrt{0.305 \cdot (d_t - CG)} \cdot \left(\frac{1}{0.43}\right) \tag{1.29}$$

In 2003, Toor and Araszewski [1.30] applied a power relationship to the dataset used by Toor [1.33] and Happer [1.19], and published the following empirical equations, along with the prediction intervals. In these equations, S is the throw distance in meters and V_V is the vehicle impact speed, in kilometers per hour. For a forward projection trajectory, the equation is as follows:

$$V_V = 8.25 \cdot S^{0.61} \pm 7.7 \text{ km/h} \tag{1.30}$$

For a <u>wrap trajectory</u>, Toor and Araszewski published the following empirical relationship:

$$V_V = 9.84 \cdot S^{0.57} \pm 5.8 \text{ km/h} \qquad \textbf{(1.31)}$$

An English unit version of these equations is included below. In these equations, S is the throw distance in feet and V_V is the vehicle impact speed in miles per hour.

For a <u>forward projection</u> trajectory:

$$V_V = 2.6 \cdot S^{0.61} \pm 4.8 \text{ mph} \qquad \textbf{(1.32)}$$

For a <u>wrap trajectory</u>:

$$V_V = 3.11 \cdot S^{0.57} \pm 3.6 \text{ mph} \qquad \textbf{(1.33)}$$

To illustrate the use of these equations, consider the staged collisions introduced earlier in this chapter and conducted by the authors of this book. In Test 2, which resulted in a wrap trajectory, the Chevrolet Malibu was traveling at approximately 40.1 mph when it first contacted the dummy. The dummy contacted the vehicle's front bumper and grille and then wrapped onto the hood and windshield in a location near the longitudinal centerline of the vehicle with a slight offset to the passenger side. The vehicle traveled 57.1 ft (17.4 m) after first contacting the dummy, and the dummy traveled 80.9 ft (24.7 m). Based on the throw distance of the dummy, Equation (1.31) would yield a range of speeds of 38.0 ± 3.6 mph (61.2 ± 5.8 km/h), a range that includes the actual speed of the vehicle.

Test 3, on the other hand, resulted in a fender vault trajectory, so Equation (1.31) would not be applicable. In this test, the vehicle was traveling at approximately 42.3 mph at the first contact with the dummy. From first contact to rest, the Malibu in this test traveled approximately 69.2 ft (21.1 m), and the test dummy traveled approximately 68.4 ft (20.8 m). If we did apply Equation (1.31), it would yield a range of speeds of 34.5 ± 3.6 mph (55.5 ± 5.8 km/h), a range that does not include the actual speed of the vehicle. This demonstrates the importance of accurately classifying the trajectory type prior to applying an empirical throw distance formula. Empirical relationships between vehicle collision speed and pedestrian throw distance generally only apply for forward projection and wrap trajectories.

In a 2021 study, Neale et al. [1.11] analyzed videos of 21 sideswipe and minor overlap pedestrian collisions. These authors proposed the following: "In a sideswipe or minimal overlap impact, the pedestrian and vehicle do not reach a common velocity. In fact, the pedestrian only reached a percentage of the vehicle's speed, which can be as little as 20%. Video analysis of the pedestrian impacts showed that a multiplier of 1.5 to 3 can be used with Toor's empirical wrap method to estimate the vehicle speed. To further refine which multiplier to use, the investigator may need to determine how much engagement has occurred between the pedestrian and the striking vehicle. If the vehicle exhibits damage consistent with more engagement, then a multiplier of 1.5 may be applicable. If evidence supports a minimal engagement with the pedestrian, a 3.1 multiplier may be more applicable." In the subject staged collision, the dummy experienced significant initial engagement with the vehicle, so these authors would propose using a 1.5 multiplier in this instance. This would yield a speed range of 51.8 ± 3.6 mph (83.3 ± 5.8 km/h), a range that overestimates and does not include the actual impact speed. The method proposed by Neale et al. is further critiqued later in this chapter. Suffice it to say for now: don't use it.

Pedestrian Injuries

While a biomechanics expert will usually be the one evaluating the specific mechanisms and forces leading to specific injuries, reconstructionists can gain insight from a cursory review of the injuries. The location and nature of the injuries can assist a reconstructionist with determining the direction the pedestrian was facing at impact, which leg was leading when the pedestrian was struck, or whether the pedestrian was standing, crouching, or laying down at impact. As an example, consider a pedestrian collision that involved a Honda Odyssey minivan striking a middle-aged female pedestrian. The pedestrian, who would have been crossing from the passenger's side of the vehicle to the driver's side, was struck by the driver's side front corner and was thrown into the oncoming lane. On-scene police photographs revealed collision damage to the driver's side front of the Odyssey, including damage to the hood, the left headlight assembly, the left front fender, the A-pillar,

the bumper fascia, the windshield, and the left side mirror. The A-pillar was dented above the side mirror, and there was a piece of scalp and hair lodged between the window trim and the A-pillar. According to the autopsy report, the pedestrian had a laceration to the back of her head that corresponded to the scalp and hair on the vehicle. The autopsy report also stated that the pedestrian had fractures of her left tibia and fibula. The location of these injuries was consistent with the Honda Odyssey striking the pedestrian when her left side was exposed to the vehicle. Further, her left leg was struck first, and then her body rotated such that the back of her head struck the A-pillar of the Odyssey, depositing the piece of scalp and hair.

Analysis of a pedestrian's gait position based on such evidence could be carried out using multibody simulation. To illustrate this, consider simulations run with the PC-Crash multibody pedestrian model. The next two figures compare the results of two simulations. **Figure 1.27** depicts the starting positions for the vehicle and the pedestrian in the simulations. **Figure 1.28** depicts the resulting head impact orientations from the two simulations. In both simulations, a 1968 Barracuda Formula S 383 strikes a pedestrian, with the left side of the pedestrian being contacted first. The initial speed of the vehicle in the simulations was 30 mph (48 km/h) with heavy braking applied just prior to the collision and continuing until the vehicle reaches its rest position. In the first simulation, the pedestrian's left leg is leading, and in the second simulation, the pedestrian's left leg is trailing. The simulation with the left leg leading results in the rear of the pedestrian's head impacting the hood of the vehicle, whereas the simulation with the left leg trailing results in the front of the pedestrian's head impacting the hood.

Figure 1.27 Simulations with the left leg versus right leg forward.

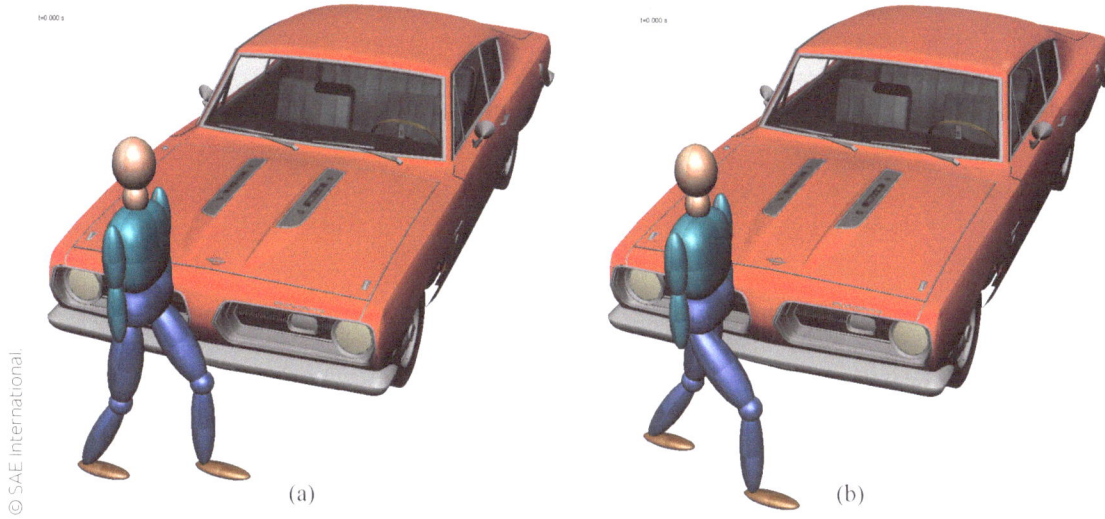

(a)   (b)

© SAE International

Figure 1.28 Simulations with impact to the back versus the front of the head.

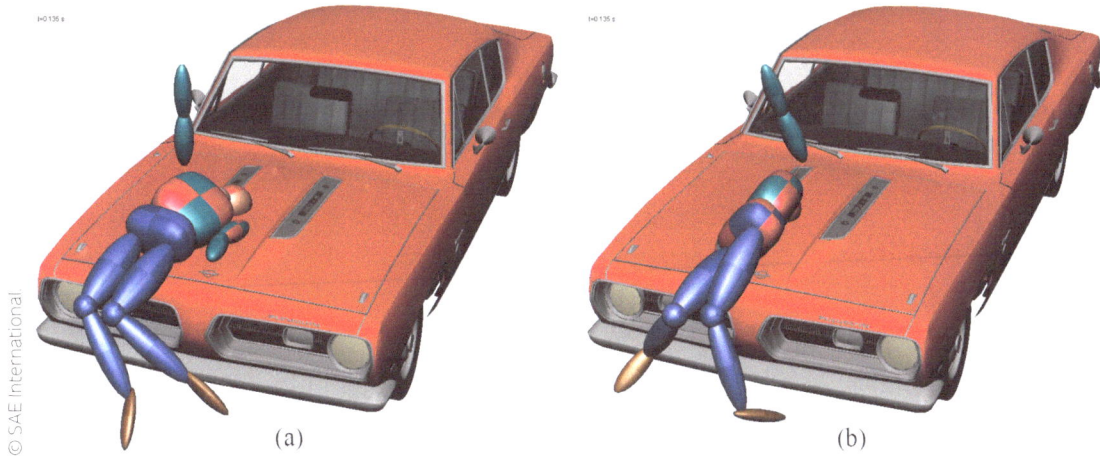

(a)   (b)

© SAE International

Pedestrian Walking and Running Speeds

To determine whether a driver could have avoided striking a pedestrian, the pre-collision speeds of both the vehicle and the pedestrian will usually be needed. The avoidance analysis will begin by determining the relative positioning of the vehicle and the pedestrian in the moments leading up to the collision. Then, the reconstructionist can determine when or where a typical or reasonable driver would have been able to perceive and respond to the pedestrian. And finally, the reconstructionist can determine whether that would have afforded the driver sufficient time to avoid (had they been attentive, for example). Sometimes, the physical evidence will enable the determination of the collision speed of the pedestrian. Or, at least, the physical evidence may enable us to determine whether the pedestrian was walking or running when they were struck, since this speed can influence how the pedestrian interacts with the vehicle. Still, a pedestrian can alter their speed and body posture in anticipation of the impact, so knowing the collision speed of the pedestrian does not necessarily tell us the speed the pedestrian was going prior to their response. Soni et al. [1.36] reported: "Accident situations were simulated with volunteers using a non-impacting methodology. Fifty one reactions from 23 volunteers of two age groups were observed. Most of the volunteers were found to run, step-back or stop in fright in a dangerous situation." More specifically, out of 51 trials, 25 pedestrians accelerated in response to the impending collision, 15 froze, 5 backed up, and 6 did not alter their speed. These results, of course, imply the pedestrians were aware of the potential of being struck—at least the ones that reacted. Unaware pedestrians would not alter their speed or body posture in response to an impending collision. In the present context, the point being made is simply that, even if there were a reconstruction technique that enabled determination of the pedestrian speed at impact, it may not be valid to assume that the pedestrian did not alter their speed prior to impact. For pedestrian collisions not captured on video, the reconstructionist will be dependent on witness statements about the pedestrian's walking speed and will likely need to consider a range of speeds.

There will not usually be tangible evidence of the pre-collision speed of the pedestrian beyond the information we might have from witnesses (unless there is video of the moments leading up to the collision). Often, the reconstructionist's assessment of the speed of the pedestrian will come from a combination of witness statements and empirical observations of pedestrian walking and running speeds reported in the technical literature [1.37]. The witness statements can help categorize the speed of the pedestrian: "He was walking." Or "She was running." If the category can be established, then typical speeds associated with that category can be selected. If there are no witnesses (or video), then the reconstructionist may need to consider a range of pedestrian speeds wide enough to encompass both walking and running. This section reviews published studies related to the speeds of pedestrians who were walking or running. There are many such studies, and the intent of this section is not to be exhaustive, but instead to highlight the factors that influence these speeds and to suggest some studies that a reconstructionist may consult in establishing a range of speeds for a pedestrian. Some of these studies were conducted within a crash reconstruction context, and others in a traffic engineering setting where walking speeds are of interest for timing pedestrian traffic signals. The walking speed studies covered in this section are reviewed chronologically.

When examining observational studies of pedestrian speeds, one consideration is whether or not the documented pedestrians were aware they were being observed. As with any study that documents human behavior, awareness of being observed can influence the characteristics of the behavior. Of course, awareness of being observed does not alter the physical limits of pedestrians, nor does it prevent the researchers from accurately characterizing the behavior of the observed pedestrians (walking, running, shuffling, etc.). Thus, studies in which the subjects are aware they are being observed are useful, but perhaps this factor could sometimes be an explanation of differences with other studies where the participants were naïve to being observed. Beyond that factor, the following factors can influence the walking or running speeds of pedestrians, and these factors could be considered when selecting a likely range of pre-collision speeds: (1) age; (2) gender; (3) weight; (4) if the pedestrian is walking alone or in a group; (5) weather conditions; (6) lighting conditions; (7) whether or not the pedestrian is walking inside or outside of a marked crosswalk; (8) signalization characteristics at the inter-section; (9) proximity of approaching traffic; and (10) whether or not the pedestrian has a disability that limits their mobility.

In a 1991 article, Thompson [1.38] reported speeds of walking and running pedestrians (unaware that they were being documented) at several intersections in Boise, Idaho. This one-and-half-page study was published in *Accident Reconstruction Journal* with little to no data reported about the methodology, weather, or traffic conditions. Thompson did not indicate whether the pedestrians were walking alone or in a group. The reported data were segmented by gender and age. For pedestrians older than age 20, male pedestrians walked on average faster than female pedestrians of the same age. Male pedestrians between 20 and 65 years of age walked at a speed of approximately 5 ft/s (1.5 m/s), on average. Male pedestrians younger than 20 years were on average mildly slower, and male pedestrians older than 65 years were signifi-cantly slower—approximately 2.7 ft/s (0.8 m/s). Female pedestrians between 20 and 65 years of age walked on average approximately 4.5 ft/s. Younger female pedestrians were mildly faster, and female pedestrians older than 65 years were slower—approximately 3.2 ft/s (1.0 m/s). There were fewer observations of running pedestrians

and (presumably because of this) fewer discernable trends. For males, running speeds varied between 6½ and 13 ft/s (2.0 to 4.0 m/s). For females, running speeds varied between 6½ and 8½ ft/s (2.0 to 2.6 m/s).

In a 1995 study, Coffin and Morrall [1.39] reported speeds of pedestrians over the age of 60 years from six field locations and a "seniors club." The field locations were in Calgary and included pedestrian-actuated midblock crosswalks, crosswalks at signalized intersections, and crosswalks at unsignalized intersections. The pedestrians at the field locations were unaware they were being observed. The pedestrians at the seniors club were aware they were being observed. The authors reported that "people over the age 60 are not a homogeneous group; they possess a range of walking speeds and mobility levels." This is a statement that could be made of any age group, and analysis of the pre-collision positions of the pedestrian will usually utilize a range of the speed of the pedestrian. In the Coffin and Morrall study, average walking speeds for the women observed at the field locations varied between 3.7 and 4.5 ft/s (1.1 to 1.4 m/s), and that for the men observed at the field locations varied between 3.9 and 4.8 ft/s (1.2 to 1.5 m/s).

In a 1996 study, Knoblauch et al. [1.40] studied walking speeds and start-up times for pedes-trians and observed that these were influenced by environmental, traffic, and pedestrian characteristics. These authors defined start-up time as the time from the onset of a walk signal until the pedestrian stepped off the curb. Sometimes, start-up times for pedestrians will be useful in a reconstruction setting, but they rarely will be a decisive variable in an avoidance analysis. Knoblauch et al. conducted field studies involving 16 crosswalks at signal-controlled intersections in Richmond, Virginia; Washington, DC; Baltimore, Maryland; and Buffalo, New York. They observed pedestrians at these intersections over the course of 8-h collection periods. In total, 7123 pedestrians were observed, including 3458 under 65 years of age and 3665 over the age of 65 years. They recorded the following information about the intersections: street width, posted speed limit, curb height, grade, number of travel lanes, signal cycle length, pedestrian signal type, street classification, crosswalk type, and channelization. Data were collected during the following weather conditions: dry, rain, and snow. The following types of pedestrians were excluded from the dataset: children under 13 years of age; pedestrians carrying children, heavy bags, or suitcases; pedestrians pushing strollers or grocery carts; pedestrians using a cane, walker, or crutches; people in wheelchairs; and pedestrians walking bikes or dogs. Pedestrians who crossed diagonally, stopped to rest, waited for traffic, entered the roadway running, or entered the roadway more than 4 ft outside of the crosswalk were also excluded. The gender of each subject and whether or not they were walking in a group were also recorded. According to Knoblauch, "a group was defined as two or more pedestrians crossing the roadway at about the same time, regardless of whether or not they were apparently friends or associates."

In Knoblauch's study, "pedestrian crossing times were measured with a hand-held, digital, electronic stopwatch. The watch was started as the subject stepped off the curb and stopped when the subject stepped up on the opposite side curb after crossing." Knoblauch noted that many site and environmental factors were statistically significant; however, he noted that "it is important to consider the relative magnitude of the differences and whether or not the differences are meaningful." He found that the mean walking speed for pedestrians under 65 years of age was 4.95 ft/s (1.5 m/s), and for pedestrians over 65 years of age, it was 4.11 ft/s (1.25 m/s). The 15th percentile speed for pedestrians under 65 years was 4.09 ft/s (1.25 m/s), and for pedestrians over 65 years, it was 3.19 ft/s (1.0 m/s). Younger males had the fastest mean walking speeds—5.11 ft/s (1.6 m/s); older females had the slowest—3.89 ft/s (1.2 m/s). Knoblauch also noted that "pedestrians who start on the Walk signal walk more slowly than those who cross on either the flashing Don't Walk or the steady Don't Walk." He found that weather conditions influenced the walking speed, with the walking speed in snow or rain being mildly higher than in dry conditions.

In another 1996 study, Bowman and Vecellio [1.41] reported pedestrian walking speeds for urban and suburban medians located on unlimited-access arterials in Atlanta, Phoenix, and Pasadena, Los Angeles. Three types of cross-sections were studied—raised medians, two-way left turn, and undivided—and pedestrian speeds were documented at both midblocks and intersections. Pedestrian were grouped into three age categories: less than 18 years old, aged 18 to 60 years old, and older than 60 years. These authors noted that "pedestrians aged 18 to 60 exhibit a significantly higher walking speed for [two-way left turn] medians for both signalized intersections and midblock locations." The mean walking speed for midblock crossings with a two-way left turn median was 1.47 m/s (4.82 ft/s) versus 1.17 m/s (3.83 ft/s) for undivided roadways. The mean walking speed for signalized intersections with the two-way left turn median was 1.46 m/s (4.79 ft/s) versus 1.19 m/s (3.90 ft/s) for the undivided roadway. These authors further reported that "the walking speed for the age group 18 to 60 is significantly higher than that of the over-60 age group for both signalized intersections and midblock locations…both age groups have significantly higher walking speeds at midblock locations than at signalized intersections." Overall, the mean speed for pedestrians in the 18–60 age group for midblock crossings was 1.41 m/s (4.63 ft/s) and for signalized intersections was 1.35 m/s (4.43 ft/s). For the over-60 age group, these values were 1.19 m/s (3.90 ft/s) and 1.03 m/s (3.38 ft/s), respectively.

In the 1999 edition of their pedestrian accident reconstruction book, Eubanks and Hill [1.13] reported walking and running speeds based on a study of nearly 3000 individuals. These pedestrians were aware they were being observed, and Eubanks and Hill reported examples of the instructions given, such as "run as fast as you can." The individuals in the study ranged in age from 17 months to 60 years. Regarding the age range from 17 to 24 months, these authors noted: "Even with specific directions on what to do, the children in this test group did whatever they wanted." They further stated: "Originally, the test data was to be obtained for walking and running conditions. However, during the test it was obvious that most children this age do not understand the term walking or running. In fact, even if the child did know the term 'walk' they would take a few steps and then start to run and then return to walking." Because of this, the authors recommended that within their test data the "85th percentile should be considered for running, the 50th percentile should be used for walking, while the 15th percentile should be applied for a dawdling child."

For younger children, it is also worth considering that children experience significant physical changes over short periods of time. As an example of how this observation could

be relevant, consider that Eubanks and Hill reported a "running" speed for 17- to 24-month-olds of 4.29 ft/s (1.3 m/s). However, they conducted additional observations at two different preschools, this time observing running speeds of nine "2-year-old" males and reporting running speeds between 5.3 and 6.4 ft/s (1.6 to 2.0 m/s), with a 50th percentile value of 5.6 ft/s (1.7 m/s). Which would be the most appropriate value for a 24-month-old? Within the dataset for "2-year-olds," the test subjects were parsed by their age in full years, and thus, these subjects could have been anywhere between 24 and 35 months old. Given this, the high end of this range is likely too high for a 24-month-old. Rapid physical changes in the children were also evident in the data reported by the authors for preschool children between 2 and 4 years. For combined male and female data, two-year-olds had a 50th percentile walking speed of 2.8 ft/s (0.9 m/s). Three-year-olds walked at approximately 3.5 ft/s (1.1 m/s), and four-year-olds at approximately 4.1 ft/s (1.2 m/s). By the age of seven years, the reported walking speeds were approximately equal to a typical adult male walking speed of approximately 5 to 5½ ft/s (1.5 to 1.7 m/s).

In 1999 and 2000, Vaughan and Bain [1.42, 1.43] published a two-part study related to the speeds and accelerations of young pedestrians at primary schools in Sydney, Australia. These pedestrians were between the ages of 5 and 17 years. The first part of this study focused on children between 5 and 11 years, and the second part added children between 12 and 17 years. These pedestrians were aware they were being observed, and they were given instructions to walk or run from a standing start over a dry, level distance of 15 m (49.2 ft). A timer was automatically started when the pedestrians broke an infrared light beam at the start, 5 m (16.4 ft) after the start, and 15 m (49.2 ft) from the start. The times for the pedestrians to cross the first 5 m (16.4 ft) and the remaining 10 m (32.8 ft) were measured. In analyzing the resulting data, the authors assumed that the pedestrians had completed their acceleration in the first 5 m (16.4 ft) and that, in the remaining 10 m (32.8 ft), they moved at a steady speed. In their first article, the authors noted that "some of the children had difficulty in coping with the tasks and occasionally a running speed was obtained, but not a walking speed. There were also a small number of exceptionally tall and/or heavy children who were excluded from the height or weight versus speed analyses because in the height or weight ranges concerned, there was only a sample of one or two." In their second article, the authors similarly noted: "During testing, we had the impression that subjects had the most difficulty in deciding just what consti-tuted 'walking' and least difficulty in 'running.' 'Jogging' was more easily determined than 'walking,' but not as readily determined as 'running.'" This highlights a potential method-ological problem for studies in which the pedestrians are aware they are being observed. Asking a pedestrian to move in a way that meets a certain description can introduce subjectivity and unrealistic variability into the resulting data.

In considering their results from their first set of testing, the authors commented: "In general terms, the results indicate that the spread of speed is wider for height and weight than it is for age: age seems to be a good basis for speed selection. If, however, a pedestrian is at an extreme end of a height or weight range for a particular age group, then it would be prudent to also take into consideration the results for height or weight (as appropriate), as well as for age." The authors nonetheless reported that linear regression analysis using the speed data yielded poor correlations with age, height, and weight. They stated that "this is consistent with the clear impression that the walking speeds obtained in testing were more variable and appeared to relate to the children's difficulty in deciding just what amounted to 'walking.' Running was not such a problem for the children. Within the running speed data, age correlated better than height, and height correlated better than weight."

Despite these results from the first article, in their second article, the authors reported empirical equations "which enable calculation of acceleration and speed ranges for pedestrians of different ages, heights and weights." Perhaps the greatest value in this pair of studies is how they illustrate the methodological problems that can arise in a study of human behavior in which humans are aware they are being watched. That is not to say that the numerical results in this study are unreliable, but certainly it is to say that studies of pedestrian behavior in which the pedestrians are not aware of being observed are preferable. In this pair of articles, the authors also emphasize that, for some reconstructions, it can be important to include the pedestrian's start-up acceleration when quantifying the time available for a driver to respond to the pedestrian.

In a study published in 2000, Smith [1.44] reported pedestrian speeds obtained by the Lawrenceville (Georgia) Police Department. The pedestrians in this study were volunteers who were aware they were involved in the study. Their ages ranged from 24 to 54 years, and they were of varying physical conditions, weights, and genders. According to Smith, "each subject was asked to walk the measured distance twice at each suggested speed of stride. The speeds of stride were described to each subject as 'casual walk,' 'quick walk,' and 'trot as if crossing a busy street.'"

These tests produced the following averages: 4.39 ft/s (1.3 m/s) for a casual walk, 6.63 ft/s (2.0 m/s) for a quick walk, and 10.11 ft/s (3.1 m/s) for a trot.

In a 2001 study, Fugger et al. [1.45] reported the behavior of pedestrians at eight intersections in Los Angeles, examining how much time elapsed between the illumination of a pedestrian walk sign and gait initiation, the pedestrians' rate of acceleration from a stop, their steady state velocity, and the number of steps required for them to reach their steady-state velocity. Fugger et al. modified the signal timing at each intersection to eliminate any all-red. This means that when the "cross traffic signal phased from green to yellow and then red, the pedestrian direction changed from red to green immediately…" Second, the pedestrian buttons were internally disabled (though they were left in place), and the pedestrian walk signal was set to automatically activate when the light became red for the cross traffic. The actions of the pedestrians were captured with a high-speed camcorder capturing video at 120 Hz. Motion tracking software was then used to track the CG motion of the pedestrians. To determine the perception-response times of the pedestrians, the authors determined the time from the onset of the walk signal to the initial movement of the pedestrian. The resulting data are included in **Table 1.3**, where the data are parsed by gender and approximate age. Within those categories, the pedestrians were further categorized by their actions. In relation to these categories, Fugger stated that the "level of anticipation was subdivided into pedestrians who were looking directly at the 'walk' signal, pedestrians who were anticipating crossing either by watching the opposing traffic signal or flow of traffic, and pedestrians who were distracted in some way." Noncompliant pedestrians were those who began crossing early. Typically, an accident reconstructionist will not know if a pedestrian was looking directly at the walk signal, anticipating the light change, distracted (in relation to the signal), compliant, or noncompliant. Fortunately, a pedestrian's perception-response time to a walk signal will usually not be a decisive variable in the causation of a crash. Still, perception-response data may be useful for some evaluations of the pre-collision relative positions of the pedestrian and the striking vehicle and for avoidance scenario calculations.

Table 1.3 Pedestrian perception-response times from Fugger et al.

	Looking straight ahead at walk signal (sec)	Anticipating light change (sec)	Distracted (sec)	Noncompliant (sec)	Compliant (sec)
Males < 55 years old	0.79 ± 0.46 (*n* = 77)	0.99 ± 0.71 (*n* = 33)	1.91 ± 1.02 (*n* = 35)	−0.41 ± 0.66 (*n* = 7)	1.11 ± 0.83 (*n* = 145)
Females < 55 years old	0.74 ± 0.38 (*n* = 59)	0.86 ± 0.49 (*n* = 26)	1.51 ± 0.85 (*n* = 21)	−0.13 ± n/a (*n* = 1)	0.94 ± 0.62 (*n* = 106)
Males > 55 years old	1.05 ± 0.66 (*n* = 70)	1.10 ± 0.79 (*n* = 10)	2.69 ± 0.95 (*n* = 7)	−0.61 ± 0.40 (*n* = 11)	1.19 ± 0.82 (*n* = 87)
Females > 55 years old	1.25 ± 0.78 (*n* = 72)	0.92 ± 0.61 (*n* = 13)	3.17 ± 2.00 (*n* = 12)	−1.03 ± 0.75 (*n* = 6)	1.44 ± 1.19 (*n* = 98)

Table 1.4 lists pedestrian speed and acceleration data reported by Fugger et al. In discussing these data, these authors concluded that "the elderly pedestrians have a slower acceleration rate and steady-state walking speed compared to the younger population." This was not universally true, though, and at Intersection 6, older pedestrians had a faster mean walking speed. Fugger et al. also reported that "the average step at which the observed pedestrians reached a constant walking velocity was 1.51 ± 0.56 steps.

Relative to the gait cycle, this infers that a steady-state walking speed is generally reached before heel strike of the stance foot." The Fugger et al. data exhibit some dependence on the intersection, suggesting that some reconstructions may warrant an intersection-specific study of pedestrian walking or running speeds. Such a study could potentially be conducted in an inconspicuous manner with the use of a camera-equipped small unmanned aerial system (sUAS, aka drone).

Table 1.4 Pedestrian speed and acceleration data from Fugger et al.

	Mean acceleration, g <55 years old	Mean acceleration, g >55 years old	Steady-state velocity, ft/s (m/s) <55 years old	Steady-state velocity, ft/s (m/s) >55 years old
Intersection 1	0.25 ± 0.20	n/a	5.15 ± 0.69 (1.57 ± 0.21)	n/a
Intersection 2	0.15 ± 0.07	n/a	4.40 ± 0.82 (1.34 ± 0.25)	n/a
Intersection 3	0.13 ± 0.06	0.11 ± 0.02	4.56 ± 0.66 (1.39 ± 0.20)	4.20 ± 0.85 (1.28 ± 0.26)
Intersection 4	0.16 ± 0.06	0.13 ± 0.04	4.00 ± 0.59 (1.22 ± 0.18)	3.61 ± 0.75 (1.10 ± 0.23)
Intersection 5	0.11 ± 0.03	0.07 ± 0.00	4.49 ± 0.69 (1.37 ± 0.21)	3.77 ± 0.49 (1.15 ± 0.15)
Intersection 6	0.14 ± 0.04	0.07 ± 0.00	4.72 ± 0.82 (1.44 ± 0.25)	5.02 ± n/a (1.53)
Intersection E5	n/a	0.09 ± 0.03	n/a	3.18 ± 0.95 (0.97 ± 0.29)
Intersection E1	n/a	0.08 ± 0.03	n/a	3.64 ± 0.95 (1.11 ± 0.29)
Intersection E2	n/a	0.09 ± 0.04	n/a	3.87 ± 0.89 (1.18 ± 0.27)
Combined Data	0.14 ± 0.09	0.08 ± 0.04	4.53 ± 0.79 (1.38 ± 0.24)	3.71 ± 0.89 (1.13 ± 0.27)

In 2001, Toor et al. [1.46] reported walking speeds of elementary school children between the ages of 5 and 14 years. The subjects in this study were unaware that their speeds were being documented. These pedestrians, who were crossing the street at ten different marked crosswalks adjacent to elementary schools, were videotaped at 30 fps with a camera hidden in a parked vehicle that was not visible from the crosswalk. Speeds were determined by analyzing this video footage. The roadways were level, and the majority of the data were collected under sunny and clear conditions. One day of data collection was conducted under rainy conditions. Data collection resulted in walking speeds for 406 children (240 males and 166 females). In addition to these 406 children, there were 53 subjects who were omitted from the dataset because they were running, jogging, or skipping across the road. Some pedestrians were also omitted because they were "bouncing a ball, walking with a bicycle, walking with a dog or pushing a scooter." Showing similar results to the Eubanks and Hill data, by the age of seven years, the reported walking speeds were approximately equal to typical adult walking speeds. The walking speeds reported by Toor et al. were mildly lower than those reported by Eubanks and Hill. Toor et al. reported that "walking speeds for both male and female pedestrians decrease when they traverse crosswalks in groups of 2 or more." Also, "young pedestrians were found to walk faster to school in the morning than when leaving school in the afternoon. Both male and female pedestrians reduced their median walking speeds in the afternoon by about 8%." The data showed "no notable difference between the walking speeds of children who were and were not carrying bags." For pedestrians observed under rainy conditions (21), more than half ran across the road and were therefore omitted from the dataset.

In 2006, Fitzpatrick, Brewer, and Turner [1.47] reported walking speed data from 42 sites from seven states (Arizona, California, Maryland, Oregon, Texas, Utah, and Washington). The sites included nine different types of pedestrian crossing treatments (half signals, Hawk beacon, midblock traffic control signal, passively activated overhead yellow flashing beacons, overhead flashing beacons active by push button, pedestrian crossing flags, high-visibility markings and signs, in-street pedestrian crossing sign, and pedestrian median refuge island). A total of 3155 pedestrians were observed in this study, 81% of whom were walking and 19% were running, both walking and running, or using some form of assistance such as skates or a bicycle. Non-walking pedestrians were not included in the analysis. There were also 107 walking pedestrians whose age could not be determined and six whose gender could not be determined. These also were not included. The 15th and 50th percentiles of the observed walking speeds by age group are included in **Table 1.5** and by age and gender in **Table 1.6**. The influence of age group is more significant in these results than the influence of gender.

Table 1.5 Walking speed by age group.

	Sample size	15th percentile Walking speed, ft/s (m/s)	50th percentile Walking speed, ft/s (m/s)
Elderly or physically disabled	15	2.75 (0.84)	3.38 (1.03)
Older (60+, but not elderly)	92	3.19 (0.97)	4.38 (1.34)
Middle (31–60)	1464	3.82 (1.16)	4.75 (1.45)
Young (19–30)	789	3.83 (1.17)	4.78 (1.46)
Teen (13–18)	76	3.79 (1.16)	4.64 (1.41)
Child (0–2)	9	3.51 (1.07)	4.37 (1.33)
All pedestrians	2445	3.82 (1.16)	4.78 (1.46)

Table 1.6 Walking speed by age and gender.

	Sample size	15th percentile Walking speed, ft/s (m/s)	50th percentile Walking speed, ft/s (m/s)
Male			
Young	1434	3.75 (1.14)	4.78 (1.46)
Old	75	3.11 (0.95)	4.19 (1.28)
All	1509	3.67 (1.12)	4.75 (1.45)
Female			
Young	890	3.79 (1.16)	4.67 (1.42)
Old	31	2.82 (0.86)	4.41 (1.34)
All	921	3.75 (1.14)	4.67 (1.42)

In 2009, Montufar et al. [1.48] reported research conducted in Winnipeg, Manitoba, Canada, to determine whether the walking speeds of pedestrians not crossing a street were different from the speeds of pedestrians crossing a street at a signalized intersection. They noted that "data were collected throughout the city [at 8 signalized intersections] as people went about their normal daily activities. None of the pedestrians in the study were aware that they were being observed." Data were collected for a total of 1792 pedestrians—1104 in the summer and 688 in the winter. Montufar et al. reported that, "in all cases the normal walking speed is less than the crossing walking speed. It also found that younger [20 to 65 years old] pedestrians walk faster than older [greater than 65 years old] pedestrians, regardless of the season and gender,

and females walk slower than males, regardless of the season and age. Furthermore, both younger and older pedestrians have a greater normal walking speed in summer than in winter," but a greater crossing walking speed in winter than in summer.

In 2010, Carson [1.49] reported the speeds of pedestrians in a crosswalk at a major sports complex. The roadway was asphalt and was straight and level. The time for the pedestrians to traverse a 39-ft distance was recorded using a digital stopwatch. Pedestrians were timed prior to the beginning of the sports event and before the crowds became so dense that a pedestrian's walking speed may have been affected by walking within crowds across the roadway. Only pedestrians who walked a straight line and were walking alone were timed. There was no precipitation falling during the documentation. Carson recorded a total of 391 measurements. The average speed for the pedestrians was 4.28 ft/s (1.3 m/s). He found no influence of ethnicity or gender, but age did have an effect. "Children younger than 11 generally moved slower than 4 ft/sec. Pedestrians in their late teens walked the fastest of any age group, clocking in at an average of 4.7 ft/sec. Adults whose age estimates ranged from 21 to 50 collectively averaged almost exactly 4.0 ft/sec. Pedestrians in the 51 to 65 age range had an average walking speed of 3.3 ft/sec, as did those estimate to be 66 years of age or older."

Carson captured additional pedestrian walking speed data over the years following the 2010 article and reported these results in 2024 [1.50]. The protocol consisted of measuring the distance the pedestrians traveled, with the time beginning when they stepped off the curb or into the measured area and stopping when their lead foot exited the measured area or went onto the curb. The time was measured using a stopwatch. Only pedestrians who were walking in a straight line were included. The age of each pedestrian was estimated by the analysts, which consisted primarily of experienced police officers. The subjects were not aware they were being observed. "Unusual" situations, such as talking on a cell phone or walking with a cane, were noted in the data. The final dataset consisted of nearly 1300 samples. Carson segmented the data based on the subjects' age in 10-year segments. He reported that the average walking speeds for females was 4.05 ft/s (1.2 m/s) and for males, it was 4.57 ft/s (1.4 m/s). For instances when the pedestrians were talking on the phone or texting, the average walking speed was 4.20 ft/s (1.3 m/s). Carson reported that "when the data for each age category was analyzed separately…and then compared it showed that walking speeds for the

0 – 10-year-old category was low (4.46 fps). Walking speeds peaked in the 11 – 20-year-old category (5.02 fps) and then steadily declined for the remainder of the age categories until it was lowest at the 71+ group (3.21 fps)."

Lu and Fernie [1.51] examined pedestrian behavior at a two-stage crossing with a center refuge island. The pedestrian signals for this crossing were timed separately on the two sides of the refuge island, such that pedestrians were expected to cross one side and then wait on the refuge island for another signal before crossing the other side. These authors compared the speeds of pedestrians who complied with the pedestrian signals to those who did not. They also examined the influence of temperature and weather on the compliance rate and on the pedestrian speeds. This study utilized a single eight-lane divided road in downtown Toronto. The pedestrians were filmed without their knowledge from a nearby rooftop. The authors reported that "a total of 484 pedestrians (78%) complied with the signal when they started the first stage crossing and a total of 135 pedestrians (22%) complied with the signal at stage 2 crossing." The average walking speed for noncompliers was 5.48 ± 0.98 ft/s (1.67 ± 0.30 m/s), and for compliers, it was 4.23 ± 0.69 ft/s (1.29 ± 0.21 m/s). Outdoor temperature and weather had a significant effect on the compliance rate, with the compliance rate decreasing in cold and snowy conditions. In terms of walking speed, compliers walked faster on cold days than on warm days—4.56 ± 0.69 ft/s (1.39 ± 0.21 m/s) versus 4.07 ± 0.43 ft/s (1.24 ± 0.13 m/s). Noncompliers also walked faster on cold days—5.64 ± 0.79 ft/s (1.72 ± 0.24 m/s) versus 5.18 ± 0.56 ft/s (1.58 ± 0.17 m/s).

In a 2013 study, Jakym, Attalla, and Kodsi [1.52] reported the crossing speeds of 242 adults crossing midblocks and having to make gap acceptance decisions. These authors noted that "none of the pedestrians involved in this study were aware of the study being conducted, nor were any of the vehicles along the roadway." The data were collected in front of a community center in Ontario. The roadway was straight and level with three through lanes in each direction. There was a single lane dedicated to the left-turning traffic between the eastbound and westbound lanes. The speed limit was 60 km/h (37.3 mph). The motion of the pedestrians was captured with high-definition video cameras at 30 fps, and speeds were determined from the video. The authors reported that "traffic on the roadway had an effect on pedestrian behavior; however, there were exceptions. For example, there were pedestrians who ran across the road when there were no vehicles in close proximity, and there were some pedestrians who casually walked across the road despite vehicles that were

in close proximity…the speed of pedestrians appeared to be affected by the lane of the approaching primary hazard vehicle. For example, the average speed of pedestrians when the primary hazard vehicle was in the near (1st) lane was 1.71 m/s (5.6 ft/s), while the average speed of pedestrians when the primary hazard vehicle was in the middle (2nd) lane was 1.83 m/s (6.0 m/s). Furthermore, the average speed of pedestrians was 1.99 m/s when the primary hazard vehicle was in the far (3rd) lane…when the gap was approximately 20 seconds or greater, the proximity of the vehicle had little effect on the crossing speed of the pedestrian. Where the gap was less than about 20 seconds, the speed of the pedestrian would be influenced by the primary hazard vehicle's proximity. As the gap decreased, pedestrians were observed, in general, to travel faster." The authors' findings related to the gap size are summarized in **Table 1.7**.

Table 1.7 Influence of gap on pedestrian speed.

	<5 sec	5–10 sec	10–15 sec	15–20 sec	>20 sec
Average, m/s (ft/s)	2.42 (7.94)	1.97 (6.46)	1.70 (5.58)	1.56 (5.12)	1.55 (5.09)
Std Dev, m/s (ft/s)	0.94 (3.08)	0.50 (1.64)	0.46 (1.51)	0.26 (0.85)	0.29 (0.95)
Min, m/s (ft/s)	1.32 (4.30)	1.02 (3.35)	0.98 (3.22)	1.14 (3.74)	0.97 (3.18)
Max, m/s (ft/s)	4.58 (15.03)	3.48 (11.42)	2.89 (9.48)	2.18 (7.15)	2.25 (7.38)
Count	14	114	75	33	60

© SAE International

For gaps less than 20 sec, Jakym et al. proposed the following empirical equation. In this equation, the gap size is in seconds and the speed is in meters per second. To convert the result to feet per second, multiply by 3.28.

$$\text{Speed} = 1.47 + 1.72e^{-0.16\cdot\text{gap}}. \tag{1.34}$$

Uncertainty Analysis

This chapter has reported certainty ranges for some of the available analysis methods and empirical datasets. There will always be uncertainty in a reconstruction, and it is common for reconstructionists to report a range on their calculations (or at least to acknowledge the uncertainty, if asked). Further discussion of this issue will be useful at this juncture. Let us begin with defining the following terms and being clear how they are used in this book and in a reconstruction setting: (1) error; (2) uncertainty; (3) accuracy; and (4) precision. The discussion in this section draws on how these terms are defined in a previous study [1.53]. Consider a scenario in which the reconstructionist is determining the speed of a vehicle depicted moving through the video footage from a stationary surveillance camera. We can begin by saying that there is a speed that the vehicle was actually traveling at any given moment it was visible to the camera. Ideally, the analysis of the video would yield vehicle speeds perfectly in alignment with the actual vehicle speed. Such an analysis would have zero error, since the error is defined as the difference between the calculated speed and the actual speed. Here, it is important to say that in a reconstruction setting, the word error is not to be equated with the term *blunder*, which the American Society for Photogrammetry and Remote Sensing (ASPRS) defines as [1.54]: "A mistake resulting from carelessness or negligence." Zero error is not achievable in the real world, and so, the next best situation would be for the error in an analysis to be known and reported. However, even this is not possible because, for a real-world collision reconstruction, the actual speed of the vehicle is not known. If it

was known, the reconstruction calculations would not be needed.

Because the error in our analysis cannot actually be measured or known, we introduce the concept of uncertainty. Quantifying the uncertainty in our calculations is our attempt to establish a range within which the true value of speed will be contained. In other words, we attempt to estimate what the error in our calculations could be based on the uncertainties inherent in the analysis. In estimating the uncertainty in the calculated value, a likely range for each of the inputs is established based on how precisely that input is or can be known. In the example of determining the speeds of a vehicle from a video, the reconstructionist might establish a range on the vehicle positions obtained from the video analysis and a range on the time between the video frames. These uncertainties in the inputs propagate to uncertainties in the calculated speed. As an example of this type of analysis, Bartlett et al. [1.55] published a study quantifying the uncertainty in measurements commonly used in crash reconstruction. Most of the data for their study were collected through measurements taken by participants at the World Reconstruction Exposition in 2000 (WREX 2000) in College Station, Texas. This study included distance measurements utilizing a 25-ft carpenter's tape measure, a flexible measuring tape, and a roller wheel. Using the flexible measuring tape, two "short" measurements of 36.06 and 38.50 ft had standard deviations of 0.017 and 0.025 ft (0.2 and 0.3 in.). Two "long" measurements of 90.60 and 91.60 ft had standard deviations of 0.061 and 0.060 ft (0.73 and 0.72 in.). The distribution of measurements was found to be normal with a mean that coincided closely with the actual measurement. Using a single-wheel roller wheel, a "short" measurement had a standard deviation of 0.081 ft (0.97 in.) and a "long" measurement had a standard deviation of 0.116 ft (1.39 in.). Using a dual-wheel roller wheel, a "short" measurement had a standard deviation of 0.076 ft (0.91 in.) and a "long" measurement had a standard deviation of 0.160 ft (1.92 in.). This study also quantified uncertainties for measurements of the radius of an arced tire mark, for measurements of an angle, for frictional drag measurements, and for vehicle crush measurements. The results from this study are, of course, specific to the measurement techniques employed in this study, but this is a good reference for those methods and also a good example of how the uncertainty in the inputs to a reconstruction could be quantified.

So, what does it mean for a reconstruction analysis to be accurate? If, after considering the uncertainty and establishing a range for the calculated value, the calculated range for a speed contains the real value of that speed, then the reconstruction is accurate. But, again, this cannot be known in the real world, other than through historical application of the analysis method to cases where the real value is known, or at least reasonably well known. This is the purpose of validation studies that compare the results produced by various analysis methods to the known values from staged collisions. These studies produce a track record for these methods, establishing the level of accuracy that would be expected from a method when applied in a real-world setting.

A reconstructionist's speed range could be stated in multiple ways. One way would be for the reconstructionist to cite a range of speeds and to say that any speed within that range is equally probable. This approach would yield a needlessly wide range for the calculated value. More likely, the reconstructionist will give a middle or mean value for the speed with a plus/minus on that value. The mean value will often be the value that the reconstructionist considers to be the most probable value, and then, the extremes of the range will be the least probable values within the range. The probability of any particular value will fall off as we move away from the mean value in either direction. The size of the reconstructionist's range for the calculated value can be seen as a measure of the precision of the analysis. The smaller the range, the greater the precision. We could always achieve 100% certainty that the real value of the speed falls within our range by establishing a wide enough range on the speed. The ideal situation, though, would be to have a method that is both accurate and precise, meaning a method that is likely to produce a range containing the real value of the speed with a range that is relatively small.

Various authors have discussed methods for quantifying the uncertainty in accident reconstruction calculations. Brach and Dunn [1.56], for instance, published a treatise covering analytical methods of uncertainty analysis in a forensic science setting. In a 1994 article, Brach covered some of the same techniques of uncertainty analysis specifically in a crash reconstruction context [1.57]. Several other treatments have appeared in the accident reconstruction literature, including Kost and Werner [1.58], Wood and O'Riordain [1.59], Tubergen [1.60], Bartlett [1.61], and Wach [1.37]. Between these sources, three methods of uncertainty analysis are often mentioned. First, a simple high–low approach can be used, where the analyst combines the high and low ends of the input ranges to produce the highest and lowest results from the formula. This approach yields an overestimate of the uncertainty since it is improbable that the actual values of the inputs would all fall at an extreme of the ranges simultaneously. Second, an analytical approach that utilizes differential calculus can be utilized. This approach involves

first taking partial derivatives of the formula with respect to each of the variables. Then, the uncertainty in the dependent variable can be calculated using the following formula [1.57]:

$$dy = \frac{\partial y}{\partial u} du + \frac{\partial y}{\partial v} dv + \cdots \tag{1.35}$$

where

dy is the uncertainty in the dependent variable y

du and dv are the uncertainties in the independent variables u and v

This equation can be extended to accommodate any number of independent variables. When calculating the uncertainty using this equation, the partial derivatives are evaluated at a nominal or reference set of values, typically the values at the middle of the range for each variable. This approach has been applied in an accident reconstruction setting [1.58], but it can become cumbersome with many of the formulas employed by crash reconstructionists. One advantage of this approach, though, is that it allows for a comparison of the relative contributions of uncertainty in each independent variable to the overall uncertainty in the dependent variable.

Third, reconstructionists have often employed a statistical technique called Monte Carlo analysis [1.57, 1.58, 1.60]. Brach [1.57] referred to this technique as "a brute force randomized simulation on a computer of a mathematical model using appropriate statistical distributions for each of the variables." Kost and Werner [1.58] noted that, in Monte Carlo simulations, "appropriate probability distributions are assigned to

the desired input parameters, and the analyses are repeatedly performed with values of the input parameters selected in accordance with the probability distributions. The results are expressed in the form of probability distributions of each of the desired output parameters, which then allows the analyst to determine the probability of the results falling within selected ranges." Wood and O'Riordain [1.59] noted that "Monte Carlo simulation methods are well established in many fields and are successfully used where non-linear relationships between variables occur." Several software packages for performing Monte Carlo simulations are commercially available, and this type of analysis can also be carried out in Microsoft Excel [1.61].

In addition to the choice of methods for quantifying uncertainty, how a reconstructionist addresses uncertainty in their analysis will also be influenced by the context in which they are carrying out a reconstruction. In a civil litigation context, the criteria for the reconstruction will typically be what is most probable—perhaps stating conclusions on a more-probable-than-not basis or to a reasonable degree of certainty or probability. In a criminal context, on the other hand, the standard is an analysis that reaches conclusions beyond a reasonable doubt. In a civil litigation context, ranges on inputs can be considered, and the reconstructionist can state that the speed of the vehicle is within some range that represents reasonable consideration of the input uncertainties. The reconstructionist could state, for example, "The speed of the vehicle was between 52 and 62 mph." In a criminal context, on the other hand, the question may be something like, "Was the driver speeding?" If the range of speed in this context was 52 to 62 mph and the speed limit was 55 mph, then the reconstructionist would not be able to say, beyond a reasonable doubt, that the driver was speeding since part of the range falls below the speed limit.

Empirical Throw Distance Equations for Fender Vaults and Sideswipes?

As this chapter has noted, the accident reconstruction literature contains empirical formulas that relate pedestrian throw distance to the collision speed of the striking vehicle. The appropriate empirical equation for analysis depends on the type of interaction the pedestrian has with the struck vehicle. Empirical throw distance equations have been developed for, and can be applied to, forward projections and wrap trajectories. However, they are not appropriate for application to roof vaults, fender vaults, pedestrian carries, or sideswipes. To illustrate the issues that arise while attempting to apply empirical throw distance equations to these impact types, consider the empirical method proposed by Neale et al. [1.11] for

narrow overlap and sideswipe pedestrian collisions (fender vaults, for example). These authors [1.11] attempted to extend empirical throw distance equations to narrow overlap and sideswipe pedestrian collisions by adding a multiplier to the throw distance equation for wrap collisions developed by Toor and Araszewski—Equation (1.31).

Neale et al. used photogrammetric methods to analyze the videos of 21 sideswipe and narrow overlap pedestrian collisions [1.11]. Thirteen of these collisions were sideswipes, and eight were narrow overlap. The vehicle speeds for the sideswipe collisions varied between 9.5 and 38.9 mph. The vehicle speeds for the minor overlap collisions varied between 7.6 and 36.8 mph. The authors observed that, "in all of the impacts analyzed, the pedestrian never achieved a common velocity with the vehicle, but rather only a portion of the vehicles speed was imparted [to] the pedestrian…" The authors plotted throw distance and vehicle speed as calculated from their video analysis. They then used Equation (1.31) to determine what speed this equation would predict based on the throw distance determined from the video analysis. Finally, they calculated a multiplier for Equation (1.31) that, if utilized, would make the formula's prediction accurate. **Figure 1.29** presents a graph that is similar to Figure 8 in the Neale et al.'s study [1.11], which plots the throw distance versus vehicle impact speed for the 21 pedestrian collisions along with a black line representing Equation (1.31). The blue diamonds are the individual pedestrian collisions.

Figure 1.29 Throw distance versus impact speed for the analyzed collisions [1.11].

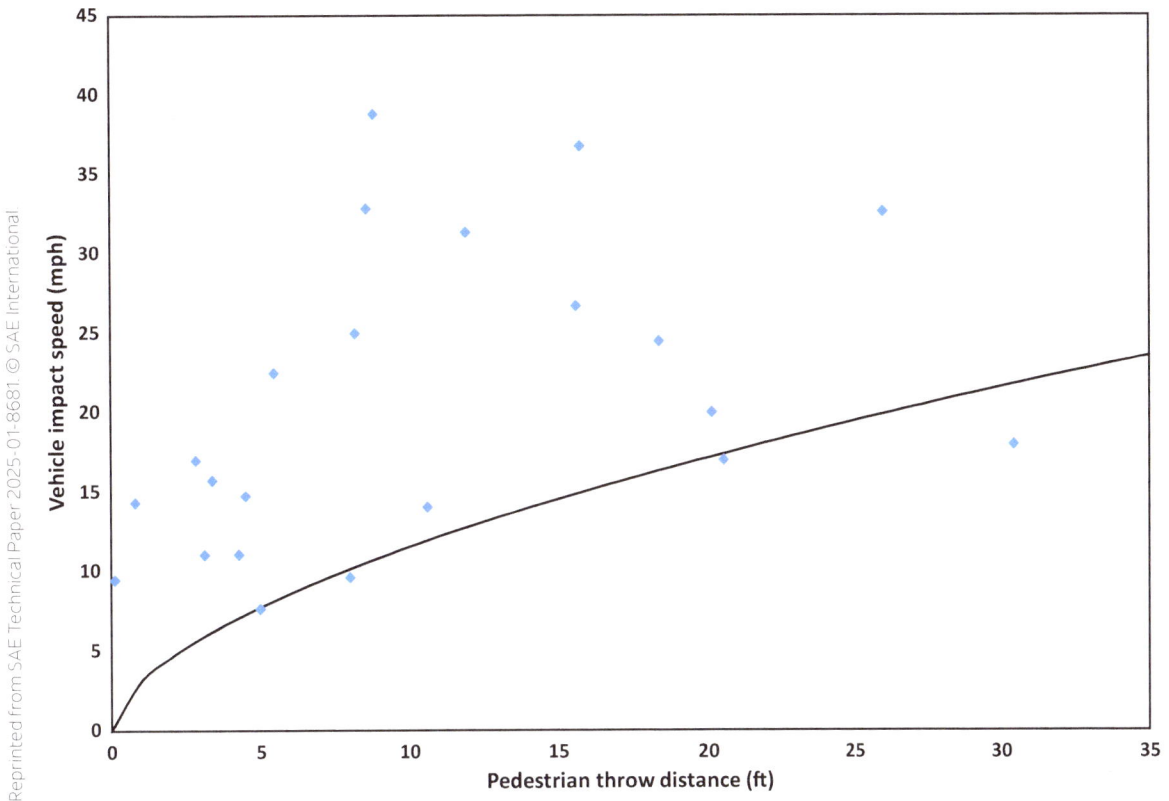

Based on this comparison, Neale et al. observed: "The comparison between calculated vehicle impact speeds and speeds from video analysis showed that the actual impact speed can be anywhere from approximately equal…to over five times more than the calculated speed [from Equation (1.31)]. This discrepancy in calculated and actual vehicle speeds showed the importance of pedestrian impact configuration and trajectory analysis in accident reconstruction, as there are potentially large differences in calculated speed values that would negatively affect a reconstruction of a similar pedestrian collision." This evaluation is both mundane and problematic since the importance of impact configuration and trajectory analysis has long been recognized in the pedestrian accident reconstruction literature and the authors did not consider the uncertainty in the empirical model represented by Equation (1.31). **Figure 1.30** is similar to **Figure 1.29,** but in this graph, instead of plotting the middle value produced by Equation (1.31), the 15th and 85th percentile values from this equation have been plotted (the two dashed black lines). The space between the two black dashed lines is the 15th to 85th percentile range from Equation (1.31). Plotting this range reveals that five of the collisions fall within the range, and another two are very close. These two would certainly be contained by the 5th and 95th percentile bands of the model.

Figure 1.30 Plotted with the 15th and 85th percentiles of the Toor and Araszewski wrap equation.

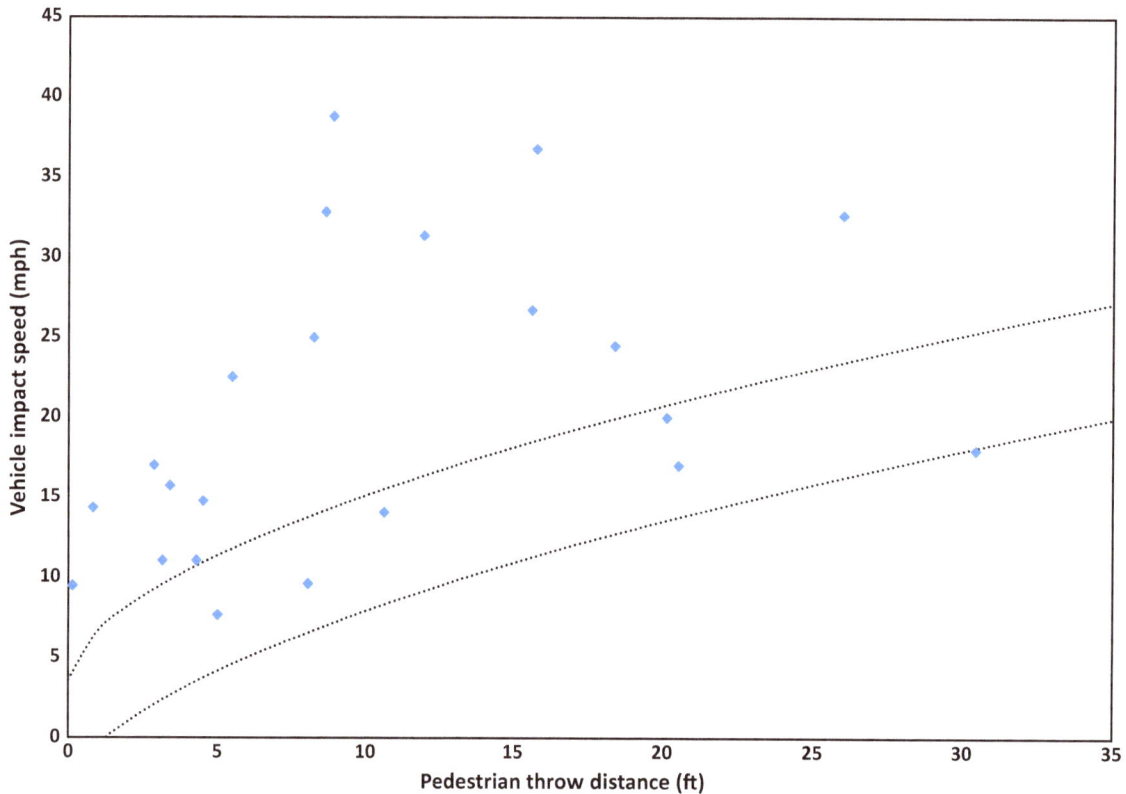

Neale et al. [1.11] did not include adequate documentation for an evaluation of the trajectory types involved in each of the collisions they analyzed, but it could be that the seven collisions adequately captured within the range of Equation (1.31) fit the assumptions of that model, whereas the other 14 collisions did not. Regardless, no multiplier needs to be applied for Equation (1.31) to adequately characterize these seven collisions. Neale et al. [1.11] continued their analysis by characterizing the interaction between the pedestrian and the vehicle in each of the collisions as *engagement* or *non-engagement*. The graph in **Figure 1.31** is similar to **Figure 1.30**, with the exception that the points for non-engagement collisions have been removed and the color of the points for engagement collisions has been changed to orange. From the examination of **Figure 1.31**, it is apparent that non-engagement collisions were further from the extents of the Toor and Araszewski wrap trajectory model. This is expected since wrap trajectories are characterized by substantial engagement between the pedestrian and the vehicle. For the remaining points on **Figure 1.31**, the 15th and 85th percentile envelope of the Toor and Araszewski model captures about half of the points.

Figure 1.31 Limited to collisions with engagement.

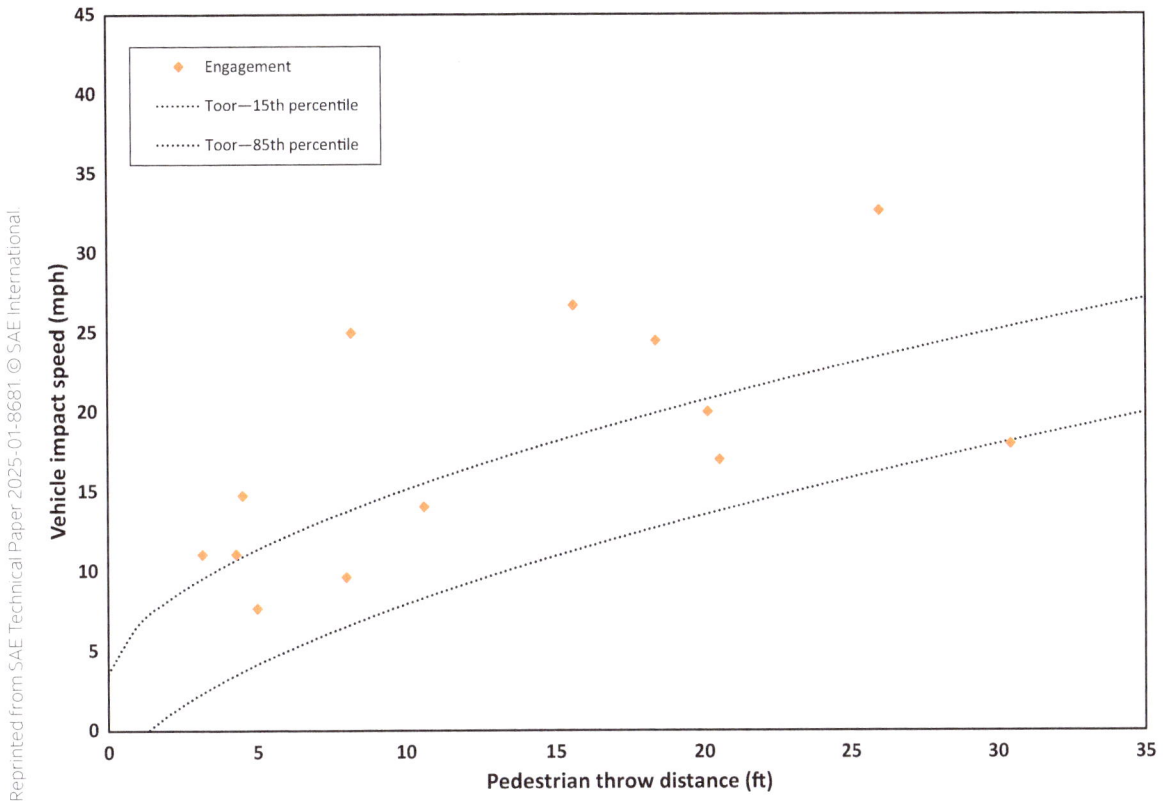

Before going further, consider several additional datasets that will help illuminate the underlying issues here. Randles et al. [1.33] studied real-world pedestrian collisions captured from a camera in a bus station clock tower "overlooking a busy downtown intersection" in Helsinki, Finland. The camera captured 15 vehicle–pedestrian collisions, and Randles et al. analyzed 13 of these using "digitizing motion analysis software to quantify the pre-impact and post-impact trajectories of both the vehicle and the pedestrian for each accident." These 13 collisions included four fender vaults, seven wrap trajectories, and two forward projections. In **Figure 1.32**, the four fender vaults from the study by Aronberg [1.32] are plotted along with the engagement collisions reported by Neale et al. [1.11], since these fender vaults are consistent with the category of collisions in Neale et al. [1.11] that had narrow overlap with good engagement. **Figure 1.32** represents these points with purple circles. These points fall outside the 15th to 85th percentile extents of the Toor and Araszewski wrap trajectory model.

Figure 1.32 Additional collisions added to Figure 3.

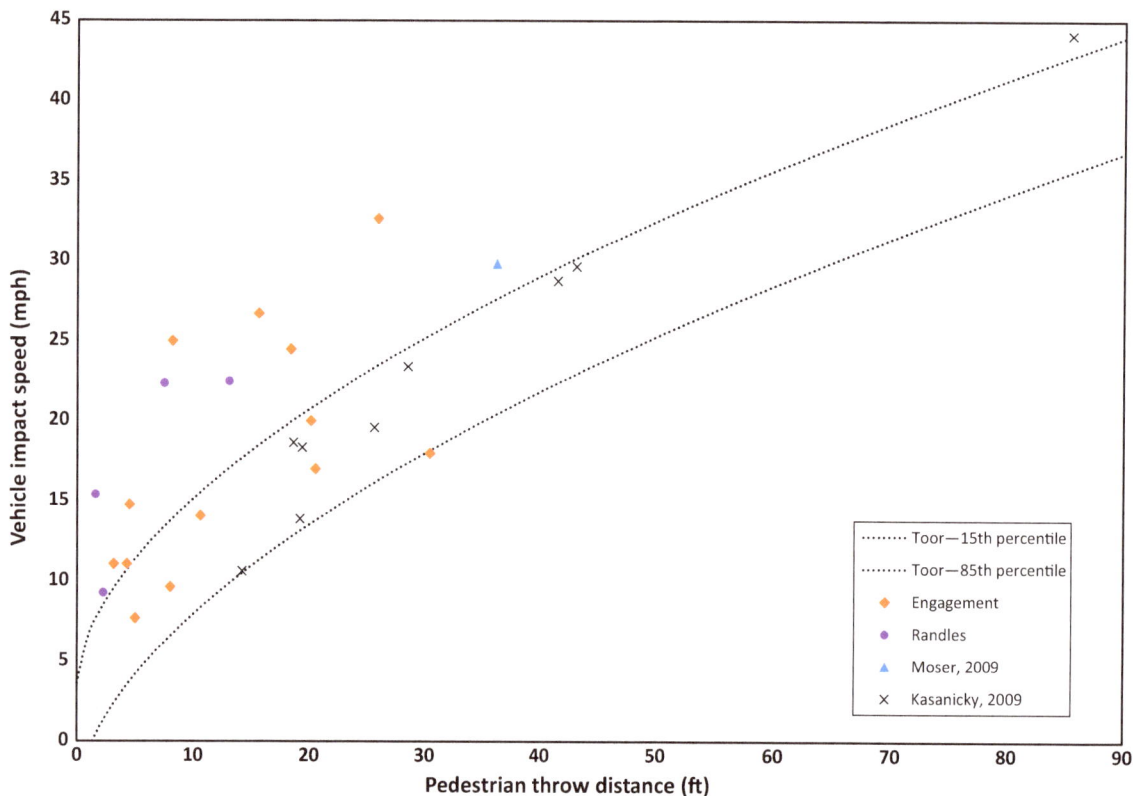

In 2009, Moser et al. [1.62] published a study at the joint ITAI-EVU Conference. This study compared a simulation using the PC-Crash pedestrian model to a real-world pedestrian collision and a crash test intended to mimic this collision. This study describes two crash tests, but only reported the speed and throw distance for one of them. This datapoint is also plotted on **Figure 1.32**, which falls outside the 15th and 85th percentile range of the Toor and Araszewski equation.

Also in 2009, Kasanicky and Kohut [1.63] published a study that presented nine successful "partial overlap" pedestrian impact tests. These authors reported thirteen total tests, but four of these were excluded from the analysis by the authors. The dataset for the remaining tests is also plotted in **Figure 1.32**. All nine points fall within the 15th and 85th percentile range of the Toor and Araszewski equation. It is interesting that despite being partial overlap collisions, Kasanicky and Kohut tests fall fully within the boundaries of the Toor and Araszewski formula. This implies greater velocity transfer to the dummy in these tests from the test vehicle than what would typically occur in a partial overlap impact configuration. This appears to be the result of the characteristics of the dummy used in these tests. This dummy is atypical for pedestrian impact tests in that it can stand without support. This indicates joints that are stiff or constrained in a way that the joints of most pedestrian dummies (or human pedestrians for that matter) are not, and this is likely to result in more significant velocity transfer to the entire dummy through these joints.

Ultimately, the reason that some of the points in **Figure 1.32** fall within the extents of the wrap trajectory model but others do not can be thought of in terms of projection efficiency, which is defined in Equation (1.5). In this equation, v_{proj} is the projection speed of the pedestrian, v_{impact} is the impact speed of the vehicle, and η_{proj} is the projection efficiency. This is a measure of the percentage of the vehicle speed that is imparted to the pedestrian.

Toor and Araszewski reported an average projection efficiency of 0.8 for wrap trajectories and an average projection efficiency of 0.95 for forward projection trajectories. The points that fall within the extents of what Equation (1.31) would predict are from collisions with projection efficiencies within the range of those characteristic of wrap trajectories. Those that fall outside the extents of Equation (1.31) have projection efficiencies lower than typical of wrap trajectories and more consistent with fender vaults. Thus, the crux of the issue is not really whether

the engagement with the pedestrian is narrow or substantial as much as it is how much of the vehicle's velocity is imparted to the pedestrian. In some instances, a collision with a pedestrian could be with the front-end corner of the vehicle, but the collision could still impart significant velocity to the pedestrian. In other instances, the collision could be with the front-end corner, but the pedestrian could fall off the side of the vehicle before a substantial portion of the vehicle's velocity was imparted to them. This could be influenced by vehicle's shape, pedestrian's velocity, pedestrian's height and posture, and pedestrian's the gait position when struck.

In our view, it is not advisable to lump fender vaults and sideswipes with wrap trajectories to try and find an empirical model that will fit both. This is what Neale et al. [1.11] attempted by introducing a multiplier to a wrap trajectory model to make it fit with fender vaults or sideswipe impacts. If one wanted to make such an attempt, one could leave the lower bound of Equation (1.31) intact and apply a multiplier to the upper bound. For example, **Figure 1.33** shows how the extents of the wrap trajectory model would change if the upper bound were multiplied by 1.7. This would encompass most of the data (excluding the sideswipes) but would produce speed ranges with too much uncertainty to be useful in most cases. For example, using this method for a pedestrian throw distance of 30 ft would result in a range of estimated vehicle speeds between approximately 18 and 43 mph. This range is likely too large to yield much insight in a reconstruction. Based on these considerations, it would be better to have a method of analysis that allows the analyst to

Figure 1.33 Plotted with the Toor and Araszewski wrap equation (15th and 1.7 × 85th percentile values).

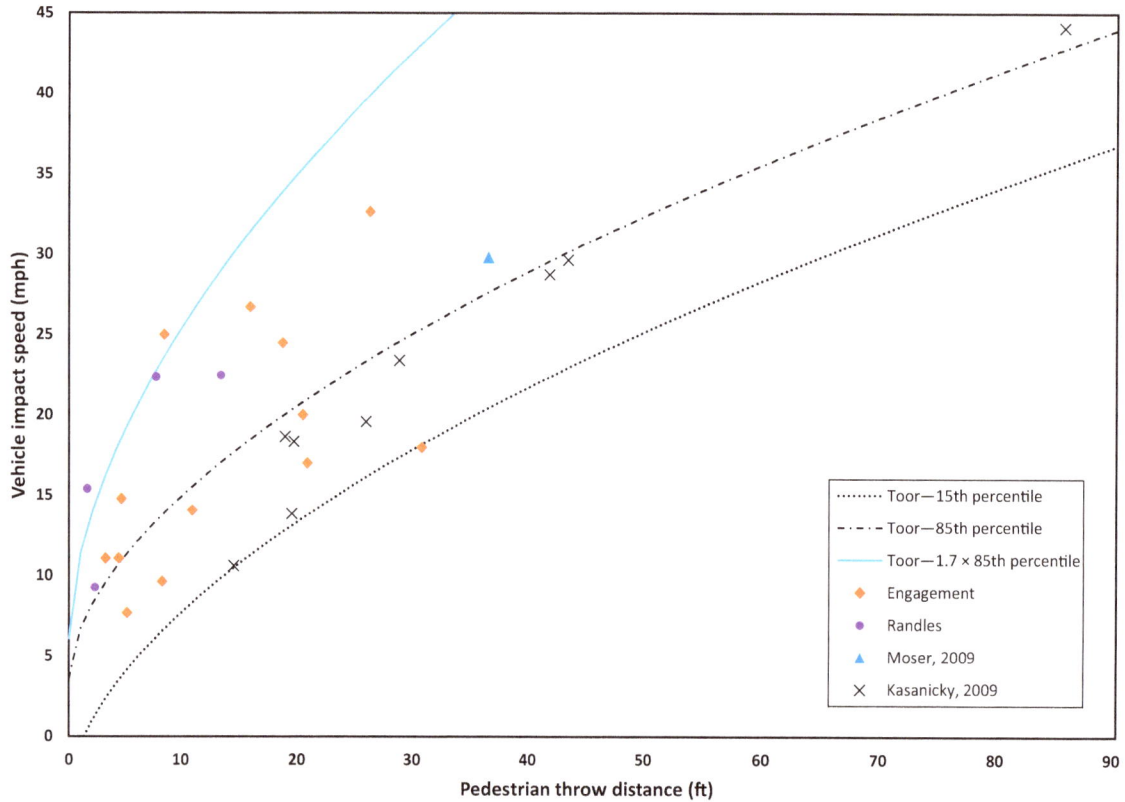

consider specific evidence (throw distance, contact points on the vehicle and the pedestrian, injury locations, pedestrian velocity, etc.) and correlate that with the speed of the vehicle. Multibody simulation, discussed in Chapter 3, is an appropriate method for this approach.

Similar criticisms could be offered in relation to Neale et al.'s [1.11] attempt to fit the Toor and Araszewski wrap trajectory model to the non-engagement collisions they studied. **Figure 1.34** shows these points (blue diamonds) plotted

in comparison to the 15th and 85th percentile extents of the Toor and Araszewski wrap trajectory model. Why attempt to fit this model to the wrap model, when a simple linear fit appears it would be adequate? There are likely too few collisions represented to say whether this pattern would hold up with a larger dataset, but there are more than adequate points to cast doubt on an attempt to morph the wrap trajectory model to fit these points.

Figure 1.34 Non-engagements plotted with the Toor and Araszewski wrap equation and a linear fit.

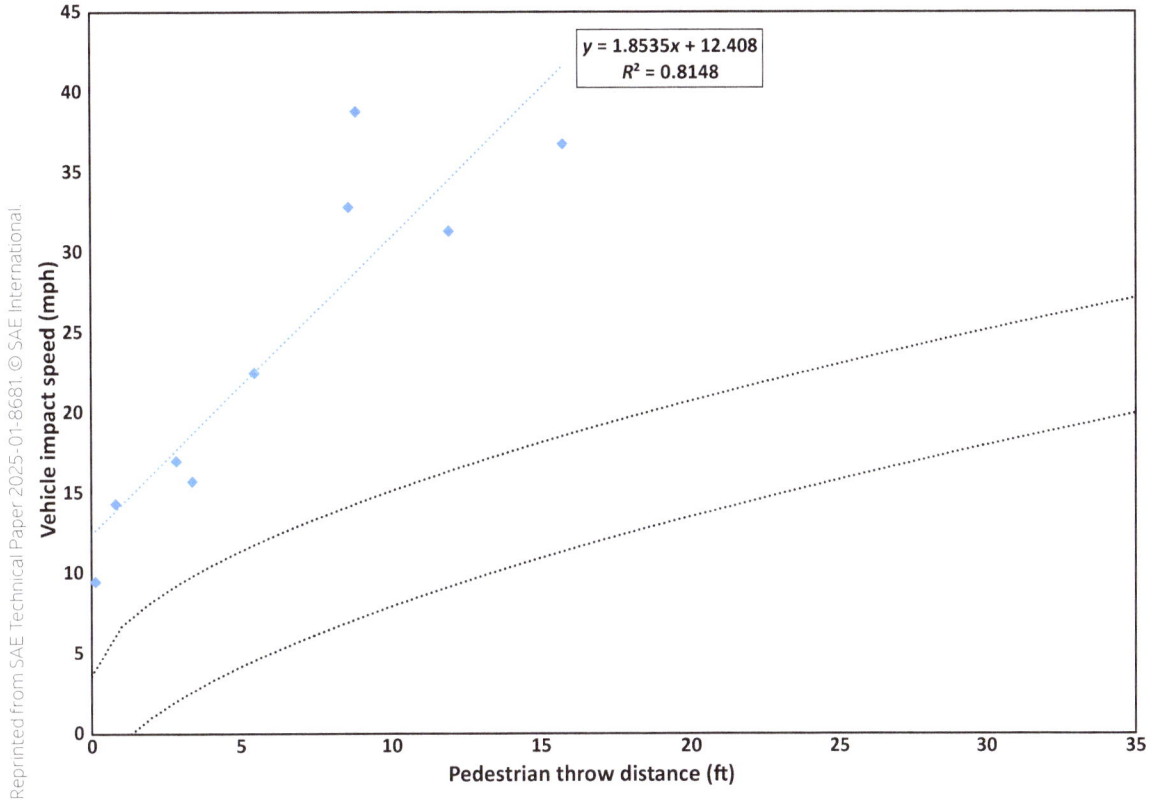

References

1.1. Wirth, J., Bonugli, E., and Freund, M., "Assessment of the Accuracy of Google Earth Imagery for Use as a Tool in Accident Reconstruction," SAE Technical Paper 2015-01-1435 (2015), doi:https://doi.org/10.4271/2015-01-1435.

1.2. Harrington, S., Teitelman, J., Rummel, E., Morse, B. et al., "Validating Google Earth Pro as a Scientific Utility for Use in Accident Reconstruction," SAE Technical Paper 2017-01-9750 (2017), doi:https://doi.org/10.4271/2017-01-9750.

1.3. Terpstra, T., Dickinson, J., and Hashemian, A., "Using Multiple Photographs and USGS LiDAR to Improve Photogrammetric Accuracy," *SAE Int. J. Trans. Safety* 6, no. 3 (2018): 193-216, doi:https://doi.org/10.4271/2018-01-0516.

1.4. Terpstra, T., Dickinson, J., and Hashemian, A., "Reconstruction of 3D Accident Sites Using USGS LiDAR, Aerial Images, and Photogrammetry," SAE Technical Paper 2019-01-0423 (2019), doi:https://doi.org/10.4271/2019-01-0423.

1.5. Severy, D. and Brink, H., "Auto-Pedestrian Collision Experiments," SAE Technical Paper 660080 (1966), doi:https://doi.org/10.4271/660080.

1.6. Fricke, L.B., "Vehicle-Pedestrian Accident Reconstruction," *Accident Reconstruction Journal* 4 (1992): 18-29.

1.7. Fricke, L.B., *Traffic Crash Reconstruction*, 2nd ed. (Evanston, IL: Northwestern University Center for Public Safety, 2010), ISBN:0-912642-3.

1.8. Collins, J.C., *Accident Reconstruction* (Springfield, IL: Charles C. Thomas Publisher, 1979), ISBN:0-398-03907-0.

Reprinted from SAE Technical Paper 2025-01-8681. © SAE International.

1.9. Ravani, B., Brougham, D., and Mason, R., "Pedestrian Post-Impact Kinematics and Injury Patterns," SAE Technical Paper 811024 (1981), doi:https://doi.org/10.4271/811024.

1.10. Stcherbatcheff, G., Tarriere, C., Duclos, P., Fayon, A. et al., "Simulation of Collisions between Pedestrians and Vehicles Using Adult and Child Dummies," SAE Technical Paper 751167 (1975), doi:https://doi.org/10.4271/751167.

1.11. Neale, W., Danaher, D., Donaldson, A., and Smith, T., "Pedestrian Impact Analysis of Side-Swipe and Minor Overlap Conditions," SAE Technical Paper 2021-01-0881 (2021), doi:https://doi.org/10.4271/2021-01-0881.

1.12. Happer, A., Araszewski, M., Toor, A., Overgaard, R. et al., "Comprehensive Analysis Method for Vehicle/Pedestrian Collisions," SAE Technical Paper 2000-01-0846 (2000), doi:https://doi.org/10.4271/2000-01-0846.

1.13. Eubanks, J.J. and Hill, P.F., *Pedestrian Accident Reconstruction and Litigation*, 2nd ed. (Tucson, AZ: Lawyers & Judges Publishing Company, Inc., 1999), ISBN:0-913875-56-2.

1.14. Reade, M. and Becker, T., *Fundamentals of Pedestrian/Cyclist Traffic Crash Reconstruction*, 1st ed., 3rd Printing (Jacksonville, FL: Institute of Police Technology and Management, 2020).

1.15. Schneider, H. and Beier, G., "Experiment and Accident: Comparison of Dummy Test Results and Real Pedestrian Accidents," SAE Technical Paper 741177 (1974), doi:https://doi.org/10.4271/741177.

1.16. Haight, W. and Eubanks, J., "Trajectory Analysis for Collisions Involving Bicycles and Automobiles," SAE Technical Paper 900368 (1990), doi:https://doi.org/10.4271/900368.

1.17. Hill, G.S., "Calculations of Vehicle Speed from Pedestrian Throw," *Impact – The Journal of the Institute of Traffic Accident Investigators* Spring (1994): 18-20.

1.18. Han, I. and Brach, R., "Throw Model for Frontal Pedestrian Collisions," SAE Technical Paper 2001-01-0898 (2001), doi:https://doi.org/10.4271/2001-01-0898.

1.19. Wood, D. and Simms, C., "Coefficient of Friction in Pedestrian Throw," *Impact – The Journal of the Institute of Traffic Accident Investigators* 9, no. 1 (2000): 12-14.

1.20. Searle, J., "The Physics of Throw Distance in Accident Reconstruction," SAE Technical Paper 930659 (1993), doi:https://doi.org/10.4271/930659.

1.21. Craig, A., "Bovington Results," *Impact – The Journal of the Institute of Traffic Accident Investigators* 8, no. 3 (1999): 19-21.

1.22. Bratten, T., "Development of a Tumble Number for Use in Accident Reconstruction," SAE Technical Paper 890859 (1989), doi:https://doi.org/10.4271/890859.

1.23. Arndt, M., "Drag Factors from Rollover Crash Testing for Crash Reconstructions," in *Proceedings of the ASME 2011 International Mechanical Engineering Congress & Exposition IMECE2011*, Denver, CO, November 11–17, 2011, IMECE2011-65537.

1.24. Rose, N. and Beauchamp, G., "Development of a Variable Deceleration Rate Approach to Rollover Crash Reconstruction," *SAE Int. J. Passeng. Cars - Mech. Syst.* 2, no. 1 (2009): 308-332, doi:https://doi.org/10.4271/2009-01-0093.

1.25. Funk, J., "An Integrated Model of Rolling and Sliding in Rollover Crashes," SAE Technical Paper 2012-01-0605 (2012), doi:https://doi.org/10.4271/2012-01-0605.

1.26. Asay, A., Carter, J., Funk, J., and Stephens, G., "Rollover Testing of a Sport Utility Vehicle (SUV) with an Inertial Measurement Unit (IMU)," SAE Technical Paper 2015-01-1475 (2015), doi:https://doi.org/10.4271/2015-01-1475.

1.27. Wood, D., "Application of a Pedestrian Impact Model to the Determination of Impact Speed," SAE Technical Paper 910814 (1991), doi:https://doi.org/10.4271/910814.

1.28. Fugger, T., Randles, B., Wobrock, J., and Eubanks, J., "Pedestrian Throw Kinematics in Forward Projection Collisions," SAE Technical Paper 2002-01-0019 (2002), doi:https://doi.org/10.4271/2002-01-0019.

1.29. Sullenberger, G., "Pedestrian Impact on Low Friction Surface," SAE Technical Paper 2014-01-0470 (2014), doi:https://doi.org/10.4271/2014-01-0470.

1.30. Toor, A. and Araszewski, M., "Theoretical vs. Empirical Solutions for Vehicle/Pedestrian Collisions," SAE Technical Paper 2003-01-0883 (2003), doi:https://doi.org/10.4271/2003-01-0883.

1.31. Searle, J., "The Trajectories of Pedestrians, Motorcycles, Motorcyclists, etc., Following a Road Accident," SAE Technical Paper 831622 (1983), doi:https://doi.org/10.4271/831622.

1.32. Aronberg, R., "Airborne Trajectory Analysis Derivation for Use in Accident Reconstruction," SAE Technical Paper 900367 (1990), doi:https://doi.org/10.4271/900367.

1.33. Randles, B., Fugger, T., Eubanks, J., and Pasanen, E., "Investigation and Analysis of Real-Life Pedestrian Collisions," SAE Technical Paper 2001-01-0171 (2001), doi:https://doi.org/10.4271/2001-01-0171.

1.34. Toor, A., Araszewski, M., Johal, R., Overgaard, R. et al., "Revision and Validation of Vehicle/Pedestrian Collision Analysis Method," SAE Technical Paper 2002-01-0550 (2002), doi:https://doi.org/10.4271/2002-01-0550.

1.35. Dettinger, J., "Methods of Improving the Reconstruction of Pedestrian Accidents – Development Differential, Impact Factor, Longitudinal Forward Trajectory, Position of Glass Splinters," *Verkehrsunfall und Fahrzeugtechnik*, Volume 12, December 1996 and Volume 1, January 1997.

1.36. Soni, A., Robert, T., Rongieras, F., and Beillas, P., "Observations on Pedestrian Pre-Crash Reactions during Simulated Accidents," *Stapp Car Crash Journal* 57 (2013): 157-183.

1.37. Wach, W. and Unarski, J., "Uncertainty Analysis of the Preimpact Phase of a Pedestrian Collision," SAE Technical Paper 2007-01-0715 (2007), doi:https://doi.org/10.4271/2007-01-0715.

1.38. Thompson, T., "Pedestrian Walking and Running Velocity Study," *Accident Reconstruction Journal* 3, no. 2 (1991): 28-29.

1.39. Coffin, A. and Morrall, J., "Walking Speeds of Elderly Pedestrians at Crosswalks," *Transportation Research Record* 1487 (1995): 63-67.

1.40. Knoblauch, R., "Field Studies of Pedestrian Walking Speed and Start-Up Time," *Transportation Research Record: Journal of the Transportation Research Board* 1538, no. 1 (1996): 27-38.

1.41. Bowman, B.L. and Vecellio, R.L., "Pedestrian Walking Speeds and Conflicts at Urban Median Locations," *Transportation Research Record* 1438 (1996): 67-73.

1.42. Vaughan, R. and Bain, J., "Acceleration and Speeds of Young Pedestrians," SAE Technical Paper 1999-01-0440 (1999), doi:https://doi.org/10.4271/1999-01-0440.

1.43. Vaughan, R. and Bain, J., "Acceleration and Speeds of Young Pedestrians: Phase II," SAE Technical Paper 2000-01-0845 (2000), doi:https://doi.org/10.4271/2000-01-0845.

1.44. Smith, J.L., "Pedestrian Velocity Trials," *Accident Reconstruction Journal* 11, no. 1 (2000): 22.

1.45. Fugger, T., Randles, B., Wobrock, J., Stein, A. et al., "Pedestrian Behavior at Signal-Controlled Crosswalks," SAE Technical Paper 2001-01-0896 (2001), doi:https://doi.org/10.4271/2001-01-0896.

1.46. Toor, A., Happer, A., Overgaard, R., and Johal, R., "Real World Walking Speeds of Young Pedestrians," SAE Technical Paper 2001-01-0897 (2001), doi:https://doi.org/10.4271/2001-01-0897.

1.47. Fitzpatrick, K., Brewer, M.A., and Turner, S., "Another Look at Pedestrian Walking Speed," *Transportation Research Record: Journal of the Transportation Research Board* 1982 (2006): 21-29.

1.48. Montufar, J., Arango, J., Porter, M., and Nakagawa, S., "Pedestrians' Normal Walking Speed and Speed When Crossing a Street," *Accident Reconstruction Journal* 19, no. 3 (2009): 11-16.

1.49. Carson, F., "Pedestrian Walking Speed in Crosswalk Study," *Accident Reconstruction Journal* 20, no. 6 (2010): 11-15.

1.50. Carson, F., "Pedestrian Age versus Crosswalk Travel Speed: Findings of a New Study," *Accident Reconstruction Journal* 34, no. 5 (2024): 18-19.

1.51. Lu, Y. and Fernie, G., "Pedestrian Behavior and Safety on a Two-Stage Crossing with a Center Refuge Island and the Effect of Winter Weather on Pedestrian Compliance Rate," *Accident Analysis and Prevention* 42 (2010): 1156-1163.

1.52. Jakym, J., Attalla, S., and Kodsi, S., "Modeling of Pedestrian Midblock Crossing Speed with Respect to Vehicle Gap Acceptance," SAE Technical Paper 2013-01-0772 (2013), doi:https://doi.org/10.4271/2013-01-0772.

1.53. Taylor, J.R., *An Introduction to Error Analysis: The Study of Uncertainties in Physical Measurements*, 2nd ed. (New York: University Science Books, 1997).

1.54. American Society for Photogrammetry and Remote Sensing (ASPRS), "ASPRS Positional Accuracy Standards for Digital Geospatial Data," Edition 2, Version 1.0, August 23, 2023.

1.55. Bartlett, W., Wright, W., Masory, O., Brach, R. et al., "Evaluating the Uncertainty in Various Measurement Tasks Common to Accident Reconstruction," SAE Technical Paper 2002-01-0546 (2002), doi:https://doi.org/10.4271/2002-01-0546.

1.56. Brach, R.M. and Dunn, P.F., *Uncertainty Analysis for Forensic Science* (Tucson, AZ: Lawyers and Judges Publishing Company, 2004), ISBN:1-930056-20-6.

1.57. Brach, R., "Uncertainty in Accident Reconstruction Calculations," SAE Technical Paper 940722 (1994), doi:https://doi.org/10.4271/940722.

1.58. Kost, G. and Werner, S., "Use of Monte Carlo Simulation Techniques in Accident Reconstruction," SAE Technical Paper 940719 (1994), doi:https://doi.org/10.4271/940719.

1.59. Wood, D. and O'Riordain, S., "Monte Carlo Simulation Methods Applied to Accident Reconstruction and Avoidance Analysis," SAE Technical Paper 940720 (1994), doi:https://doi.org/10.4271/940720.

1.60. Tubergen, R., "The Technique of Uncertainty Analysis as Applied to the Momentum Equation for Accident Reconstruction," SAE Technical Paper 950135 (1995), doi:https://doi.org/10.4271/950135.

1.61. Bartlett, W., "Conducting Monte Carlo Analysis with Spreadsheet Programs," SAE Technical Paper 2003-01-0487 (2003), doi:https://doi.org/10.4271/2003-01-0487.

1.62. Moser, A., Steffan, H., and Strzeletz, R., "Movement of the Human Body versus Dummy after the Collision," in *Proceedings of the 1st Joint ITAI-EVU Conference, 18th EVU Conference, 9th ITAI Conference*, Hinckley, UK, 2009, 87-105.

1.63. Kasanicky, G. and Kohut, P., "New Partial Overlap Pedestrian Impact Tests," in *Proceedings of the 1st Joint ITAI-EVU Conference, 18th EVU Conference, 9th ITAI Conference*, Hinckley, UK, 2009, 107-126.

Chapter 02

Emergency Braking for Late-Model Vehicles

In some instances, physical evidence will be insufficient to pinpoint precisely where on the road a vehicle collided with a pedestrian. Late-model vehicles with ABS may not leave discernable tire marks under heavy braking, and while it is possible that one of the pedestrian's shoes could scuff the road, this evidence is usually not recognized or documented by on-scene investigators. Witnesses and the involved driver may express uncertainty about where the pedestrian was prior to the collision, and usually, debris from the collision will provide too little certainty and precision in locating the collision location, particularly if the question being posed is whether or not a pedestrian crossed within a designated crosswalk. In these instances, the final resting positions of the vehicle and the pedestrian may be known. Based on these rest positions, a range can sometimes be established for the area of impact by establishing compatibility between the

throw distance of the pedestrian and the braking distance of the vehicle.

To illustrate this, consider a staged collision conducted by the authors, which was introduced in Chapter 1. This test involved a 2007 Chevrolet Malibu impacting a full-sized Rescue Randy pedestrian dummy. The Chevrolet was driven into the impact, and the driver applied aggressive braking from a speed of 49 mph (78.9 km/h) prior to impact. The dummy was placed in the center of the test fixture to align with the center of the test vehicle. At contact with the dummy, the vehicle was traveling at approximately 40.1 mph (64.5 km/h). The dummy contacted the front bumper and grille of the vehicle and then wrapped onto the hood and windshield in a location near the longitudinal centerline of the vehicle with a slight offset to the passenger side. Figure 1.5 is an evidence diagram for this test. The Chevrolet traveled

57.1 ft (17.4 m) after first contacting the dummy, and the dummy traveled 80.9 ft (24.7 m). The average deceleration of the Chevrolet post-collision was approximately 0.94g.

In Chapter 1, the empirical throw distance versus speed equation published by Toor and Araszewski [2.1] for wrap trajectories was applied to this test—Equation (1.31). In this equation, S is the throw distance, in meters and V_V is the vehicle impact speed, in kilometers per hour.

$$V_V = 9.84S^{0.57} \pm 5.8 \text{ km / h}$$

Using the throw distance of 80.9 ft (24.7 m), this equation yielded a range of speeds of 38.0 ± 3.6 mph (61.2 ± 5.8 km/h), a range that includes the actual impact speed of the vehicle. While this is a useful result, the application of Equation (1.31) depends on knowing the location of the collision, so that the throw distance can be measured. If the impact location is not known, additional data can be brought into the assessment—in particular, the likely post-collision deceleration of the Chevrolet and the difference between the post-collision travel distance of the Chevrolet and the pedestrian. This difference will be identified with the variable S_{diff} and in this instance, this distance was 23.8 ft (7.3 m). The throw distance S will be the unknown braking distance of the Chevrolet, S_{brake} added to S_{diff}. For this example, assume that the driver was aggressively braking over the entire distance from impact to rest. For this test, this was the case, but for real-world crashes, this may or may not be known. With this assumption in place, V_V is given by Equation (2.1). In order for there to be compatibility between the braking distance and the throw distance, Equation (2.1) will produce the same result as Equation (1.31):

$$V_V = \sqrt{2f_{brake}gS_{brake}} \tag{2.1}$$

To find a range of speeds and impact locations where these two equations produce compatible results, braking distances between 0 and 100 ft were considered and decelerations between 0.94 ± 0.10g. Monte Carlo simulation was utilized in Microsoft Excel to generate possible combinations of the inputs and outputs of these formulas, and the uncertainty in Equation (1.31) was considered. Approximately 10,000 iterations were run, and 815 of these produced speeds for Equations (1.31) and (2.1) that were within 0.5 mph of each other. The 15th and 85th percentile range of the throw distance and speeds for these combinations were 51.3 to 97.1 ft (17.4 to 29.6 m) and 28.4 and 44.5 mph (45.7 and 71.6 km/h), respectively. Both of these ranges contained the actual values. In this instance, it could also be assumed that the collision had to have occurred upstream of where the dummy's sunglasses came to rest. Thus, the throw distance had to be at least 65.9 ft (20.1 m). This additional constraint would reduce the size of the range on throw distance to 65.9 to 97.1 ft (20.1 to 29.6 m).

It would have to be determined on a case-by-case basis whether or not this analysis yields an adequately precise range to reach meaningful conclusions. From this case study, though, it is apparent that assessing the location of the collision can depend on a prior assessment of the post-collision deceleration of the striking vehicle. This assessment will require the reconstructionist to estimate the deceleration the striking vehicle was capable of producing, to establish when the driver of the striking vehicle began braking, and to assess how much of the braking capability the driver was utilizing.

Maximum Decelerations

In assessing the maximum deceleration achievable by the striking vehicle, the following factors could be influential: (1) the vehicle type (sedan, SUV, van, or pickup, for example); (2) the vehicle model year; (3) the type of tires on the vehicle; (4) the presence or absence of ABS; (5) the presence or absence of brake assist; (6) the surface type (asphalt, concrete, or gravel, for example); and (7) the surface conditions (i.e., wet or dry, new or deteriorated, clear or contaminated). This assessment could be made by testing an exemplar vehicle or by referencing empirical data in the technical literature.

Vehicles without ABS

In a study published in 1990, Wallingford, Greenlees, and Christoffersen [2.2] reported tire–roadway friction coefficients for concrete and asphalt surfaces. These authors stated, "the primary purpose of this paper is to quantitatively assess differences which may exist between performance type tires often utilized on law enforcement vehicles and typical production type passenger car tires relative to tire-roadway friction." The testing reported in this article consisted of 430 test runs utilizing four 1989 model year sedans (two front-wheel drive and two rear-wheel drive) and six sets of tires (bias ply, standard production radial, and

performance radial) with skidding occurring on both concrete and asphalt surfaces. The test vehicles weighed between 3000 and 4000 lb, and they were not equipped with ABS. Initial speeds of the vehicles were nominally 30 or 50 mph. Vehicle speed was measured using radar, and the distance used for calculation of the coefficients of friction was from the initial application of the brake pedal to the stopped position of the vehicle (not just the skid distance). Thus, the reported decelerations included the build-up phase of the deceleration. These authors reported that "the performance tires achieved slightly higher coefficients than both the standard production radials and the bias-ply tires." Further, "the difference between concrete and asphalt was found to be quite small…the contrast between front wheel drive and rear wheel drive was more telling…the two front wheel drive vehicles showed a higher average friction coefficient as compared to rear wheel drive vehicles. The difference was more pronounced on the asphalt surface." Overall, the authors reported averages of the decelerations between 0.75 and 0.77, with lows and highs ranging from 0.61 to 0.93.

Heinrichs et al. [2.3] reported 252 skid-to-stop tests from initial speeds of 20, 40, 60, and 80 km/h (12.4, 24.9, 37.3, and 49.7 mph) using three different radial tire grades (economy, touring, and performance) with a single 1991 Honda Accord EX-R on a single traveled asphalt road surface in "fair to good condition and without defects." The test vehicle was modified to remove the ABS. The tires used on the test vehicle were initially new, but they were subjected to 600 km (373 miles) of combined city and highway driving prior to the testing. Heinrichs reported the following formula for the skidding coefficient of friction, in which s_{skid} is the skid distance:

$$\mu_{skid} = 0.7043 + 0.2453e^{-0.3124s_{skid}}$$ (2.2)

Bartlett [2.4] summarized 17 studies, including References [2.2, 2.3], related to the maximum deceleration (or coefficient of friction) for passenger vehicles without ABS under locked-wheel skidding on dry pavement. The underlying studies were for a timeframe between the late 1980s to the mid-2000s. He concluded, "Evaluating the reported friction values from multiple different sources using a variety of vehicles, drivers, surfaces, measuring tools, and analysis techniques gives average value in the 0.76g range with standard deviations of 0.04 to 0.07g. Given an average of 0.76 ± 0.06g, 95% of all cases can be expected to fall between 0.64g and 0.88g."

Vehicles with ABS

In another study, Bartlett [2.5] reported decelerations for maximal braking on dry pavement with ABS of 0.821 ± 0.067, an increase of about 8% due to the presence of ABS. Bartlett's study also noted "a two-tailed Student's t test of published data from skid tests on concrete and asphalt roadways showed that those two data sets are statistically indistinguishable." Bartlett also reported maximal braking tests from approximately 40 mph (64.4 km/h) with four test vehicles—a sedan, a large SUV, a cargo van, and a high-performance vehicle. Each vehicle was tested with and without ABS on a concrete roadway and on a gravel roadway. Bartlett found that "on average, ABS-drag-factors on concrete were 15% higher than locked-wheel values. In contrast, the ABS-drag factors on gravel were an average of 16% lower." Based on skid test data provided by Baker Materials Engineering, Bartlett reported longitudinal friction coefficients of 0.580 ± 0.074 for braking on gravel without ABS and of 0.483 ± 0.066 for braking on gravel with ABS. Based on his own testing, Bartlett reported an average locked-wheel deceleration of 0.623g (SD = 0.058g) for gravel and an average ABS deceleration of 0.495g (SD = 0.064g).

Leiss [2.6] reported tests with a 2012 BMW 328i sedan, on both dry and wet asphalt, using three different tire types—summer (high-performance), all season, and winter. He conducted tests with and without ABS, and all of his tests had a nominal initial speed of 45 mph. Leiss reported a total of 72 tests. With ABS on dry asphalt, the summer tire produced an average deceleration of 0.98g (six tests), with a range of 0.971 to 0.998g. Without ABS on dry asphalt, the same tire produced an average deceleration of 0.70g (six tests) with a range of 0.677 to 0.727g. With ABS on wet asphalt, it produced an average deceleration of 0.95g (six tests) with a range of 0.927 to 0.967g. Without ABS on wet asphalt, it produced an average deceleration of 0.68g (six tests) with a range of 0.657 to 0.700g.

With ABS on dry asphalt, the all-season tire produced an average deceleration of 0.88g (six tests), with a range of 0.868 to 0.889g. Without ABS on dry asphalt, it produced an average deceleration of 0.73g (six tests) with a range of 0.713 to 0.749g. With ABS on wet asphalt, it produced an average deceleration of 0.85g (six tests) with a range of 0.823 to 0.869g. Without ABS on wet asphalt, it produced an average deceleration of 0.61g (six tests) with a range of 0.595 to 0.612g. With ABS on dry asphalt, the winter tire produced an average deceleration of 0.79g (six tests), with a range of 0.767 to 0.798g. Without ABS on dry asphalt, it produced an average deceleration of 0.83g (six tests) with a range of 0.802 to 0.835g. With ABS on wet asphalt, it produced an average deceleration of 0.72g (six tests) with a range of 0.698 to 0.770g. Without ABS on wet asphalt, it produced an average deceleration of 0.54g (six tests) with a range of 0.488 to 0.580g.

In a 2023 study, Miller et al. [2.7] reported maximum decelerations for 20 late-model passenger vehicles (13 cars, 3 vans, and 4 SUVs) equipped with ABS on asphalt and concrete. Eighteen tests were conducted for each vehicle from a target speed of 40 mph (65 km/h). The testing program resulted in 436 tests, and the ABS was activated in all tests. The authors reported: "Overall, we found that late-model ABS-equipped vehicles can decelerate at average levels that vary from about 0.871g to 1.081g across both surfaces, and that deceleration levels were on average about 0.042g higher on asphalt than on concrete. We also found that the passenger cars decelerated about 0.087g higher than the vans and SUVs." Overall, the deceleration on asphalt was 0.987 ± 0.064g, and the deceleration on concrete was 0.943 ± 0.065g. The decelerations of the cars were on average 0.087g higher than those of the other vehicle types. These authors reported an average brake lag of 196 ms, with a 95th percentile range of 109 to 351 ms.

Influence of Surface Debris

Debris on a roadway surface can influence the maximum achievable deceleration. In a 2006 study, Hamernik et al. [2.8] examined the influence of gravel on a road surface. They conducted 96 skid-to-stop tests on various surfaces utilizing four test vehicles—a light truck, a sedan, a sports car, and a SUV—at two nominal test speeds—20 and 40 mph (32.2 and 64.4 km/h). Each vehicle was tested with and without ABS. The baseline road surface was a dry, asphalt-paved frontage road. **Figure 2.1** is a photograph from this study showing an example of the gravel contamination that the authors tested. This study reported an approximately 0.1g reduction in average deceleration due to the presence of the gravel on the surface. There was variability in this reduction between tests, and when the ABS was active, the range of reduction was 5 to 20%. When the ABS was deactivated, the range of reduction was 13 to 20%.

Figure 2.1 Photograph depicting gravel contamination level on road surface during the Hamernik et al. testing.

Reprinted from Reference [2.8]. © SAE International.

Gravel is a common roadway contaminant, but there are others that could also be relevant in specific cases. Lambourn and Viner [2.9] examined the influence of diesel oil, engine oil, clay, and coarse sand on an asphalt surface and on a concrete surface. Hall and Painter [2.10] examined friction reductions from fine-grained earth contaminants present on a road. Meyers and Austin [2.11] examined friction reductions on dry pavement from sanding applications.

Utilization of Deceleration Capability

Another question that might arise would be the extent to which an individual driver utilized the available deceleration capability of their vehicle. If the involved vehicle does not have ABS and the vehicle deposits skid marks on the roadway, then the reconstructionist can be certain that the full braking capabilities were utilized. On the other hand, if the vehicle has ABS and no skid marks are visible on the roadway, then the reconstructionist may not know the severity of the braking. Of course, pre-crash event data downloaded from the vehicle could illuminate the braking severity, if they are available.

Ising et al. observed that, in accident reconstruction, "there has traditionally been an assumption that full braking is occurring upon completion of the mechanical brake lag. This assumption is challenged by a growing body of research indicating a concurrent driver-related delay between brake application and full braking" [2.12]. In other words, when responding to a developing hazard, drivers do not necessarily utilize the full braking capabilities of their vehicle, and even when they do, they do not necessarily move immediately to maximum braking. Similarly, Lee observed that "a driver

does not normally initiate full-power braking as soon as he sees that a lead vehicle is braking… Rather, the driver most likely adjusts his braking on the basis of his assessment of the urgency of the situation" [2.13]. Prynne and Martin noted that "accident studies have shown that the full braking capability of vehicles is not often used in emergencies…Emergency braking, therefore, is often a two-stage process with drivers rapidly depressing the brake pedal to the normal limit of depression (about a third of the full range available) and then depressing the pedal further to some lower position after they have thought about the situation" [2.14].

Ising et al. examined this issue through analysis of data from Mazzae et al. [2.15]. The Mazzae study included both dry and wet road testing. The dry road testing utilized 104 male and 88 female drivers between the ages of 25 and 55 years. The wet road testing utilized 26 males and 27 females. Test subjects were told that the study was to assess steering and speed maintenance in typical driving conditions. They drove several laps on a track, passing through a simulated intersection several times. Initially, real vehicles were situated at this intersection as if waiting to travel across the road on which the subjects were driving. Between the third and fourth laps, the real vehicles were replaced with foam replicas, and as the subjects approached the intersection on their fourth lap, the foam vehicle to the right was towed rapidly into their path, blocking half of the lane. Ising analyzed Mazzae's data and observed that "there is a driver-related braking delay occurring after initial brake application and prior to full emergency braking. In the first phase of brake application, lasting approximately 0.3 seconds, the vehicle reaches a moderate deceleration (about 0.4g). Thereafter, the vehicle deceleration profile is dependent on [time-to-intersection (TTI)] with those drivers that did ultimately apply full braking taking approximately 0.4 to 0.8 seconds longer to do so." Ising concluded his article with this statement related to crash avoidance analysis: "Finally, while the calculated braking profile can be applied when full braking is known to have been applied, this study showed that many drivers never achieved full, or even moderate, braking in this lateral incursion scenario."

Another factor that could come into play in assessing how much of the available deceleration was utilized by a driver is the presence or absence of brake assist on the vehicle. As an example, the owner's manual for a 2015 Chevrolet Impala LTZ states that the brake assist system "uses the stability system hydraulic brake control module to supplement the power brake system under conditions where the driver has quickly and forcefully applied the brake pedal in an attempt to quickly stop or slow down the vehicle. The stability system hydraulic brake control module increases brake pressure at each corner of the vehicle until the ABS activates." In a 1998 study, Yoshida, Sugitani et al. from Toyota [2.16] discussed the development of a brake assist system. These researchers ran experiments in which both male and female drivers ranging in age from 18 to 70 years were confronted with an obstacle unexpectedly emerging into their path to invoke a panic response. They noted that "the size of obstacle was set large enough so that the drivers could recognize that braking operations were absolutely necessary…The vehicle speed was set at 50 km/h, and the time allowance for braking to avoid the obstacle was set to 2 seconds." These experiments led to the conclusion that 47% of the drivers failed to apply the brakes with sufficient force to generate skidding or activate the ABS. However, the authors discovered that drivers who apply sufficient pedal force, and those who do not,

"apply the same pedal force for the first 0.05 seconds of the stop," which is important for the development of brake assist. Further, when the authors compared the panic braking responses, they could be distinguished from braking responses during normal driving. The brake assist system they developed could "recognize occurrence of a panic situation when the pedal travel speed has exceeded a given value."

AEB

When evaluating a collision, accident reconstructionists will often evaluate the nature and timing of the involved drivers' emergency responses (braking or steering). The proliferation of EDRs and dash- or windshield-mounted cameras have provided a clearer picture of these driver responses. In addition, the proliferation of AEB and other driver assistance systems on newer vehicles introduces a new wrinkle in the evaluation of driver responses. The vehicle itself may now be capable of responding, and so, accident reconstructionists will need to consider the degree to which they can distinguish the response of the driver from the response of the vehicle. This could be an important detail in a reconstruction, since intervention by the vehicle could indicate a lack of response, or an untimely response, by the driver.

AEB is typically paired with FCW, such that the first intervention by a vehicle will be a visual, audible, and/or haptic warning that aims to prompt driver intervention to avoid a collision. Haptic warnings could include vibrating the seat cushion, tugging on the seatbelt, or a haptic brake application. If the driver does not respond to this warning (or does not respond quickly enough), the AEB system will intervene with braking. The AEB response varies from one manufacturer to another. Some studies have noted a two-phase AEB response, with a mild to moderate braking phase preceding full emergency braking or, alternatively, a short build-up of heavy braking followed by a phase where the braking level is tuned to the available space. If the human driver at any point responds with braking, the AEB response will terminate, and the system will surrender control back to the driver. That said, vehicles equipped with AEB are typically also equipped with brake assist that can preload the braking system and amplify the braking response of the driver.

For a vehicle equipped with FCW and AEB, the following events could occur:

1. FCW activates, the driver responds by braking, and the collision is avoided.

2. FCW activates, AEB activates, the driver responds by braking, AEB surrenders control to the driver, and the collision is avoided.

3. FCW activates, AEB activates, the driver never responds, and the collision is avoided.

4. FCW activates, AEB activates, the driver never responds, and the collision occurs.

5. FCW activates, AEB activates, the driver responds by braking, and the collision occurs.

6. FCW activates, the driver responds by braking, and the collision occurs.

Accident reconstructionists will not encounter the first three types, which are instances in which the collision is avoided. However, they could encounter any of the event types where a collision occurs. Ideally, event data acquired from the vehicle will indicate to the reconstructionist what systems on the vehicle engaged, how they engaged, and when the driver's response occurred relative to the engagement of these systems.

Typically, AEB systems utilize a radar sensor and/or one or more cameras on the front of the vehicle to sense objects in or approaching the path of the vehicle. An AEB system can be evaluated/examined with the following questions:

- How does the system respond in a particular scenario?
- How consistent is the system's response?
- How effective is the system's response? Was the object avoided?
- At what distance from an object ahead does the system engage?
- When autonomous braking becomes necessary, what deceleration level is generated?

The Insurance Institute for Highway Safety (IIHS) tests AEB systems on passenger vehicles, and they have made their test data publicly available on their TechData site. In this testing, the subject vehicle approaches a stationary target vehicle mock-up at a nominal speed of either 20 or 40 km/h (12.5 or 25 mph). The testing is conducted on dry asphalt. The subject vehicle starts 150 to 200 m away, and as it approaches the stationary vehicle mock-up, the AEB system is allowed to respond without any driver intervention. IIHS also performs tests to evaluate the effectiveness of passenger vehicle AEB systems in avoiding pedestrian impacts. The testing protocol is frequently updated. As of January 2024, the protocol calls for three different test types: (1) an adult pedestrian crossing a street on a path perpendicular to the travel line of a vehicle at night in the dark; (2) a child pedestrian crossing a street from behind an obstruction on a path perpendicular to the travel line of a vehicle during the day; (3) and an adult pedestrian near the edge of a road on a path parallel to the travel path of a vehicle at night. Ratings are based on the test vehicle's ability to avoid or mitigate pedestrian dummy collisions at 20 and 40 km/h (perpendicular path scenarios) and at 40 and 60 km/h (parallel path scenario). Scenarios 1 and 2 are tested with a pedestrian target speed of 5 ± 0.2 km/h, while scenario 3 is tested with a stationary pedestrian. Each test vehicle is equipped with instrumentation to measure speed, longitudinal and lateral acceleration, longitudinal and lateral position, yaw rate, impact time, and FCW activation. Based on the performance of the vehicle through the series of tests, IIHS assigns a rating of poor, marginal, acceptable, or good. The IIHS dataset could be useful when a reconstructionist needs to assess the role and actions of an AEB system for a crash.

In a 2019 study, Miholjcic et al. [2.17] used the IIHS's data to examine the performance of AEB systems on model year 2013 to 2018 passenger vehicles. The dataset represented 184 vehicles from 31 manufacturers. Miholjcic et al. observed that across the model years analyzed, AEB systems showed improvements in their ability to avoid the stationary vehicle mock-up, with higher success rates in the newer-year models. Illustratively, in the 2013 to 2014 model year timeframe, 50 to 60% of the tests resulted in successful avoidance from 20 km/h and 10 to 15% from 40 km/h. For the 2018 model year, these percentages had increased to a 100% avoidance success rate at 20 km/h and a more than 80% success rate at 40 km/h. The authors also observed that all of the tested vehicles exhibited a two-phase braking response with a short initial low-deceleration phase followed by a longer high-deceleration phase. Beyond that similarity, there was variability in the response times and the braking severity from one manufacturer to another.

A question that designers of AEB systems have to consider: What is the earliest (perhaps defined in terms of time to collision [TTC] or in terms of distance) that the human driver will tolerate the AEB system intervening? The pre-2016 Toyotas were largely unsuccessful at avoiding the target vehicle mock-up from 25 mph, whereas 2016 and newer Toyotas were generally successful. This improvement was, in part, achieved through an earlier response by the AEB system, with braking beginning on average 2.2 sec prior to an impact (approximately 81 ft) as opposed to 0.66 sec (approximately 24 ft). This is presumably a tolerable difference, but as the vehicle speed increases, the need to respond sooner will at some point become intolerable to the human driver, and perhaps inappropriate, given the uncertainty that develops related to whether or not a collision will occur. In addition to braking sooner, the post-2016 Toyotas generated a higher deceleration during the first phase of AEB braking (0.12g versus 0.24g).

In a 2020 study, Siddiqui et al. [2.18] used the IIHS data to examine the effectiveness of AEB systems that include pedestrian detection and response capabilities (P-AEB). At the time of this study, the IIHS had performed testing on 11 such vehicles—4 from model year 2018 and 7 from model year 2019. The dataset included 332 tests. Siddiqui et al. observed wide variation in FCW and AEB performance across manufacturers. The IIHS testing for systems with P-AEB involved the test vehicle approaching a mock pedestrian at speeds in three different configurations: (1) an adult pedestrian crossing perpendicular to travel path of the test vehicle with no visual obstructions with the test vehicle approaching at 20 and 40 km/h; (2) a child pedestrian entering the path of the test vehicle from behind an obstruction with the test vehicle approaching at 20 and 40 km/h; and (3) an adult pedestrian traveling near the edge of the road parallel to the travel direction of the test vehicle with the test vehicle approaching at speeds of 40 and 60 km/h. The mock pedestrians move at approximately 5 km/h, representing an adult walking speed or a child running speed.

Based on their analysis of the IIHS data, Siddiqui made the following observation that will be important to accident reconstructionists using the IIHS data: "IIHS tabulated data for FCW and AEB engagement times was independently verified for each test. In many instances, we discovered AEB engagement times that required some adjustment. The IIHS test protocol states that 'the point at which the vehicle longitudinal deceleration reaches 0.5 m/s^2 is considered the start of autonomous emergency braking.' The FCW driver alert strategy for some vehicles includes a momentary haptic brake application. In tests where this FCW haptic brake application exceeded the IIHS threshold of 0.5 m/s^2, IIHS reported the haptic brake application time as the AEB engagement time, although the true application of hard automatic emergency braking did not occur until later."

In a 2021 study, Vandiver and Anderson [2.19] reported 17 tests that examined the performance of the Ford Pre-Collision Assist and AEB systems on a 2020 Ford Explorer (alert sensitivity set to normal). The Pre-Collision Assist feature is similar to what other manufacturers refer to as FCW. In this testing, the Explorer was driven at a stationary mock vehicle or stationary pedestrian dummy. Of the 17 tests, 11 were conducted with the vehicle mock-up and 6 with the pedestrian dummy. Data were collected related to the collision avoidance systems on the Ford. In some of the tests, the driver intervened by steering to avoid the target during the avoidance system's response, and in others, the driver did not

intervene. These authors concluded that the AEB system on the Ford could bring the vehicle to a stop without contacting the target from a speed slightly greater than 20 mph. The average deceleration generated by the system varied, but in this testing the highest average deceleration was 0.56g. In two tests from higher speeds (55 and 60 mph), the test driver engaged the brakes heavily prior to any intervention from the AEB system. In the 55-mph test, collision was successfully avoided without any intervention from the AEB system. In the 60-mph test, collision was not avoided. The AEB system engaged even though the test driver was already braking, but the system did not increase the deceleration above what the driver was already generating.

In 2022, Miholjcic et al. [2.20] evaluated the results of 2374 pedestrian AEB tests performed by IIHS of vehicles of model year 2018 to 2020. The analysis included tests where the subject vehicle approached the collision area with an adult-sized dummy stationary on the roadway approximately 25% of the way into the lane of travel, an adult-sized dummy travelling perpendicularly into the roadway at 5 km/h, with the projected point of impact located 25% of the way into the lane, and a child-sized dummy travelling perpendicularly into the roadway at 5 km/h, with the project point of impact located midway into the lane of travel. In the last test, the dummy was initially visually obstructed by vehicles parked adjacent to the roadway. The subject vehicles were tested at 20 and 40 km/h for tests with moving pedestrian dummies and at 40 and 60 km/h for tests with stationary pedestrian dummies. The tests were evaluated based on whether the vehicle impacted the dummy, the clearance between the vehicle and the pedestrian (for successful avoidance tests), and vehicle deceleration during AEB activation. The authors concluded that overall, 66% of the tests resulted in successfully avoiding the pedestrian dummies. Vehicles had higher avoidance rates (83%) at 20 km/h than they had at 40 km/h and above (57%). The authors also noted that most vehicle manufacturers had varying success rates across models, and some vehicle models responded inconsistently to others of the same year, make, and model.

A 2023 paper by Harrington and Martin [2.21] detailed their testing of a 2014 Subaru Forester equipped with the North American Generation 1 Eyesight System. Whereas most vehicle manufacturers use a combination of cameras and radar, the Eyesight system relies on two forward-facing stereo cameras attached to the windshield to evaluate the forward roadway. During the testing, the authors drove the instrumented vehicle toward a custom-built foam SUV-shaped stationary vehicle target at speeds between 6 and 57 mph. In each tests, the authors evaluated the TTC when the vehicle's FCW activated and the TTC when the vehicle's AEB activated. The authors concluded that the vehicle successfully stopped and avoided a collision at speeds below 35 mph, but was unable to avoid collisions at speeds above 40 mph. The average deceleration of the AEB phase increased linearly with speed up to 40 mph.

In the end, the role played by an AEB system in any particular pedestrian collision will need to be evaluated on a case-by-case basis. The evaluation will need to consider the specifics of the AEB system on the involved vehicle. It is also important to note that AEB systems are not limited to passenger vehicles. As sensor, processing, and actuation technology and hardware become widespread and affordable, heavy truck manufacturers have also developed and incorporated safety systems with AEB.

At the time of this writing, Volvo trucks, Daimler trucks (Mercedes-Benz and Detroit), Isuzu trucks, Navistar/International Trucks, Hino, WABCO, Bendix, and others offer collision mitigation systems that will automatically apply collision imminent braking. Many of these systems will also detect and apply AEB for pedestrians.

References

2.1. Toor, A. and Araszewski, M., "Theoretical vs. Empirical Solutions for Vehicle/Pedestrian Collisions," SAE Technical Paper 2003-01-0883 (2003), doi:https://doi.org/10.4271/2003-01-0883.

2.2. Wallingford, J.G., Greenlees, B., and Christoffersen, S., "Tire-Roadway Friction Coefficients on Concrete and Asphalt Surfaces Applicable for Accident Reconstruction," SAE Technical Paper 900103 (1990), doi:https://doi.org/10.4271/900103.

2.3. Heinrichs, B.E. et al., "Vehicle Speed Affects Both Pre-Skid Braking Kinematics and Average Tire/Roadway Friction," *Accident Analysis and Prevention* 36 (2004): 829-840, doi:10.1016/j.app.2003.08.002.

2.4. Bartlett, W., "Skidding Friction: A Review of Recent Research," Unpublished.

2.5. Bartlett, W. and Wright, W., "Braking on Dry Pavement and Gravel With and Without ABS," SAE Technical Paper 2010-01-0066 (2010), doi:https://doi.org/10.4271/2010-01-0066.

2.6. Leiss, P.J., Becker, S., and Derian, G., "Tire Friction Comparison of Three Tire Types," SAE Technical Paper 2013-01-0783 (2013), doi:https://doi.org/10.4271/2013-01-0783.

2.7. Miller, I., King, D., Wilkinson, C., and Siegmund, G.P., "Decelerations for Vehicles with Anti-lock Brake Systems (ABS) on Dry Asphalt and Concrete Road Surfaces," SAE Technical Paper 2023-01-0616 (2023), doi:https://doi.org/10.4271/2023-01-0616.

2.8. Hamernik, J., Paster, E., Wittekind, D., Tholl, B. et al., "Quantifying the Effects of Surface Debris on Vehicle Deceleration Rate and Anti-lock Brake Systems," SAE Technical Paper 2006-01-1676 (2006), doi:https://doi.org/10.4271/2006-01-1676.

2.9. Lambourn, R.F. and Viner, H.E., "Friction Tests on Contaminated Road Surfaces," Project Report PPR0073, TRL Limited, January 2006.

2.10. Hall, G.J. and Painter, J., "Pavement Friction Reduction due to Fine-Grained Earth Contaminants," SAE Technical Paper 2007-01-0736 (2007), doi:https://doi.org/10.4271/2007-01-0736.

2.11. Meyers, D.R. and Austin, T.P., "Dry Pavement Friction Reductions due to Sanding Applications," SAE Technical Paper 2012-01-0603 (2012), doi:https://doi.org/10.4271/2012-01-0603.

2.12. Ising, K.W., Droll, J.A., Kroeker, S.G., D'Addario, P.M. et al., "Driver-Related Delay in Emergency Responses to a Laterally Incurring Hazard," in *Proceedings of the Human Factors and Ergonomics Society 56th Annual Meeting*, Boston, MA, 2012.

2.13. Lee, D.N., "A Theory of Visual Control of Braking Based on Information about Time-to-Collision," *Perception* 5 (1976): 437-459.

2.14. Prynne, K. and Martin, P., "Braking Behaviour in Emergencies," SAE Technical Paper 950969 (1995), doi:https://doi.org/10.4271/950969.

2.15. Mazzae, E.N., Barickman, F.S., Forkenbrock, G., and Baldwin, G.H.S., "NHTSA Light Vehicle Antilock Brake System Research Program Task 5.2/5.3: Test Track Examination of Drivers' Collision Avoidance Behavior Using Conventional and Antilock Brakes," DOT HS 809 561, National Highway Traffic Safety Administration, 2003.

2.16. Yoshida, H., Sugitani, T., Ohta, M., Kizaki, J. et al., "Development of the Brake Assist System," SAE Technical Paper 980601 (1998), doi:https://doi.org/10.4271/980601.

2.17. Miholjcic, D., Fabbroni, M., and Robinson, R., "A Study of the Performance of Automatic Emergency Braking (AEB) Systems Equipped on Passenger Vehicles for Model Years 2013 to 2018," SAE Technical Paper 2019-01-0416 (2019), doi:https://doi.org/10.4271/2019-01-0416.

2.18. Siddiqui, O., Famiglietti, N., Nguyen, B., Hoang, R. et al., "Empirical Study of the Braking Performance of Pedestrian Autonomous Emergency Braking (P-AEB)," SAE Technical Paper 2020-01-0878 (2020), doi:https://doi.org/10.4271/2020-01-0878.

2.19. Vandiver, W. and Anderson, R., "Performance of the Ford Pre-Collision Assist with Automatic Emergency Braking System in Instrumented Tests," SAE Technical Paper 2021-01-0894 (2021), doi:https://doi.org/10.4271/2021-01-0894.

2.20. Miholjcic, D., Erazo, F., and Polak, A., "A Study of the Performance of Automatic Emergency Braking Systems When Presented with Pedestrian Targets," SAE Technical Paper 2022-01-0824 (2022), doi:https://doi.org/10.4271/2022-01-0824.

2.21. Harrington, S. and Martin, N., "An Evaluation of the Automatic Emergency Braking and Forward Collision Warning System in a 2014 Subaru Forester," SAE Technical Paper 2023-01-0621 (2023), doi:https://doi.org/10.4271/2023-01-0621.

Chapter 03

Simulation of Pedestrian Collisions

Nathan Rose, Connor Smith, and Neal Carter

Introduction

PC-Crash, which is a widely used crash simulation software package, includes a pedestrian model for simulating pedestrian collisions. This model is a multibody system consisting of a series of rigid bodies representing various body parts (the head, torso, and pelvis, for example) connected with joints. The user can specify the weight and height of the pedestrian, the initial position and orientation of each body part (within the constraints imposed by the joints), and the initial velocity of the pedestrian. The software will sense the contact between the pedestrian's body parts and the vehicle or the ground in the simulation and calculate the forces and resulting motion for the pedestrian. This model is particularly useful when other methods are not feasible. For example, empirical throw distance equations are only applicable to forward projections and wrap trajectories, not to fender

vaults, roof vaults, or sideswipe pedestrian collisions. This chapter describes application of the PC-Crash pedestrian model, demonstrating that the pedestrian model enables the reconstructionist to account for the physical evidence from a pedestrian collision and to determine the set of inputs (vehicle impact speed, pedestrian speed, gait position, etc.) that are most consistent with this physical evidence (vehicle and pedestrian rest positions, pedestrian-to-vehicle contact points, vehicle damage, and pedestrian injury locations). The PC-Crash pedestrian model is well validated, and this chapter is based largely on an extensive peer-reviewed study that the authors published through SAE [3.1]. The remainder of this section makes note of prior literature related to this model that the reader may want to consult. We found that, while some of these studies are helpful for demonstrating the validity of the model, there is inadequate guidance given related to inputs for the model.

The remainder of the chapter describes our validation of the model, which was accomplished by simulating staged collisions from the literature. The reader will leave this chapter with an understanding of appropriate inputs for the PC-Crash pedestrian model and with a framework for understanding appropriate inputs for similar models in other simulation software packages.

In a 1999 study [3.2], Moser et al. compared a simulation with the PC-Crash pedestrian model to a crash test involving a car and a pedestrian dummy. In this crash test, "the dummy was hit partially with the right front part of a VW Polo. The impact speed was approx. 54 km/h (33.5 mph) and the dummy was only hit on the left leg. Due to the impact forces the dummy was rotated and hit the right A-pillar of the car with the torso." Thus, this study was reporting on a narrow overlap pedestrian collision simulation. Moser et al. reported that "a good correlation between the crash tests and the simulation results in general was found. Especially the total post impact travel of the pedestrian as well as contact locations, where the pedestrian hit the car, were predicted in the simulation runs…The pedestrian model showed accurate results for the post impact movement of the pedestrian. Using the impact locations, where the pedestrian had contact with the car, the impact velocity for the vehicle could have been calculated in a range of ±5 km/h [±3.1 mph]." Friction and restitution inputs for the pedestrian model were not reported in this study.

In 2000, Moser et al. published a second study related to the PC-Crash pedestrian model [3.3]. The authors compared PC-Crash simulations to three crash tests. They presented a visual comparison between the crash tests and the simulations, and the simulations exhibited

agreement with the crash tests in terms of pedestrian motion and contact points on the vehicle. In running these simulations, the authors used a ground friction coefficient for the pedestrian of 0.600, a value that is now the default friction coefficient for the model. The authors ran additional simulations with six vehicle shapes to determine the pedestrian throw distances predicted by PC-Crash. The simulated throw distances were compared to the predictions of throw distance formulas and models from the literature. Moser et al. reported that the simulations showed general agreement with these models/formulas but noted that "depending on the shape of the vehicle different throwing distances result in the simulation." This influence of the vehicle shape is not accounted for in the throw distance models/formulas, and this is an advantage of conducting analysis using PC-Crash, or with a similar multibody model. Moser et al. concluded that "the different comparisons show that the PC-Crash pedestrian model gives very good estimates for the total pedestrian trajectory. The influence of the vehicle shape on the pedestrian kinematics and total trajectory can be taken into account easily."

In 2009, Moser et al. [3.4] published a study of the PC-Crash pedestrian model at the joint ITAI-EVU Conference. The authors noted that the PC-Crash pedestrian multibody model has the capability to incorporate and consider the shape of the striking vehicle, the detailed interaction between this shape and the pedestrian's body, braking of the striking vehicle, and the pedestrian-to-ground friction. This study compared a simulation using the PC-Crash pedestrian model to a real-world pedestrian collision and a crash test intended to mimic this collision. The authors reported that "the simulation in PC-Crash produced acceptable results for the damage locations and the initial contact phase. It produced very good results for the overall kinematics and throwing distance." The authors emphasized the importance of using an accurate vehicle model in the simulation to achieve accurate modeling of the interaction and contact points between the pedestrian and the vehicle. They also noted the usefulness of the PC-Crash pedestrian model for evaluating the influence of the various input parameters on the pedestrian motion.

Wach and Unarski [3.5] used the PC-Crash multibody model to simulate the motion of a person falling from height in a stairwell. They reported that several hundred simulations were run using a detailed computer model of the stairwell and the PC-Crash multibody. The goal was to determine from where the person had fallen and under what conditions (i.e., an accident fall or being pushed). Simulations were sought that agreed with the known physical evidence, including the rest position of the person. Wach and Unarski noted that they were unable to use the simulations to "categorically indicate either the victim's involuntary falling over or the fall being forced by a third party." They were able to determine the area where the fall originated.

Richardson et al. [3.6] used PC-Crash to examine the influence of the vehicle shape and impact location on the throw distance of a pedestrian. These authors ran 19 simulations for each of eight different vehicle shapes, varying the impact location across the front end of the vehicle from one simulation to another. Vehicle impact speeds ranged between 20 and 80 km/h (12.4 to 49.7 mph). Richardson observed that "the location where the pedestrian has engaged with the vehicle can and does significantly influence the throw distance (and projection) and subsequent

impact speed analysis." Richardson set the pedestrian-to-vehicle friction coefficient to 0.2 and the pedestrian-to-ground friction coefficient to 0.8 or 1.2. As will be demonstrated later in this chapter, this pedestrian-to-ground friction input is not within a reasonable range, and Richardson's numerical results should not be utilized.

Piloto et al. [3.7] used the PC-Crash multibody model to study pedestrian forward projection collisions. They compared the results of PC-Crash simulations to real-world data and to results from other models and analysis methods. These authors did not present their inputs for the multibody model, such as the friction and restitution parameters. They also used an integration time step of 5 ms, even though the PC-Crash technical manual states that calculations with the multibody model should be run with an integration time step of 1 ms or less. Based on the images in the paper, they also used unrealistic pre-impact pedestrian positioning.

Characteristics of the PC-Crash Pedestrian Model

Figure 3.1 depicts the default pedestrian model in PC-Crash within the multibody system dialog box. This model, which has the file name *Pedestrian 20171127.mbdef*, has a default height of 72.244 in. (1.8 m) and a weight of 176.4 lb (80 kg). The user can resize the model using the System Properties dialog box or the Body Data dialog box. According to the PC-Crash operating and technical manual, "in the case of multibodies representing humans, the stature should not be defined [in the System Properties dialog box]. For example, the height in this dialog box depends on position (i.e., a person bending over or sitting would have a lower 'height' than a standing person). For humans, the stature should be defined using the 'Change Body Data' button." **Figure 3.2** shows an example of three pedestrian multibodies with weights of 250 lb (113.4 kg), 176.4 lb (80 kg), and 100 lb (45.4 kg).

These weights were changed in the Body Data dialog box (as recommended), and PC-Crash resized the ellipsoids of the pedestrian in accordance with the entered weight.

Figure 3.1 PC-Crash pedestrian multibody.

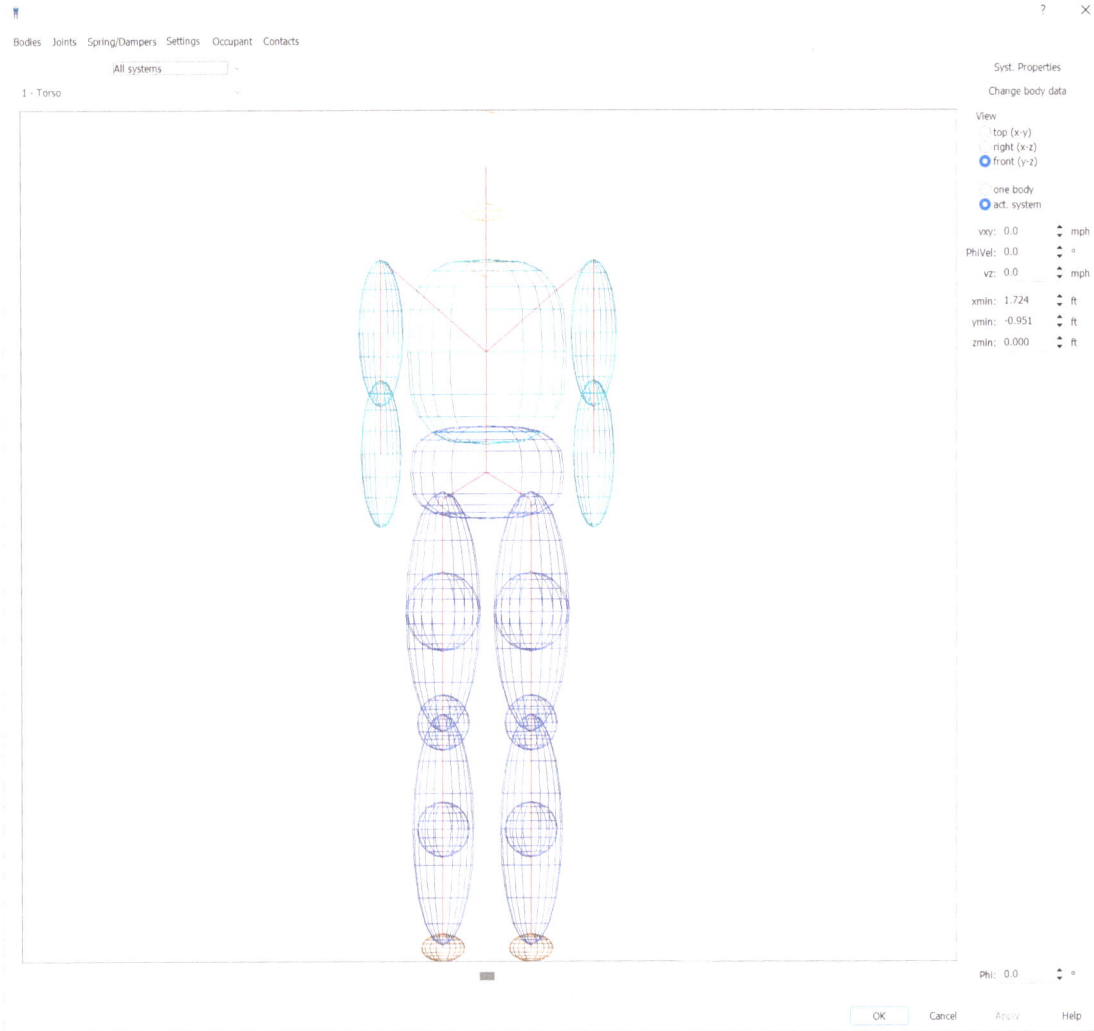

Figure 3.2 Pedestrian parameters changed in the Body Data dialog box.

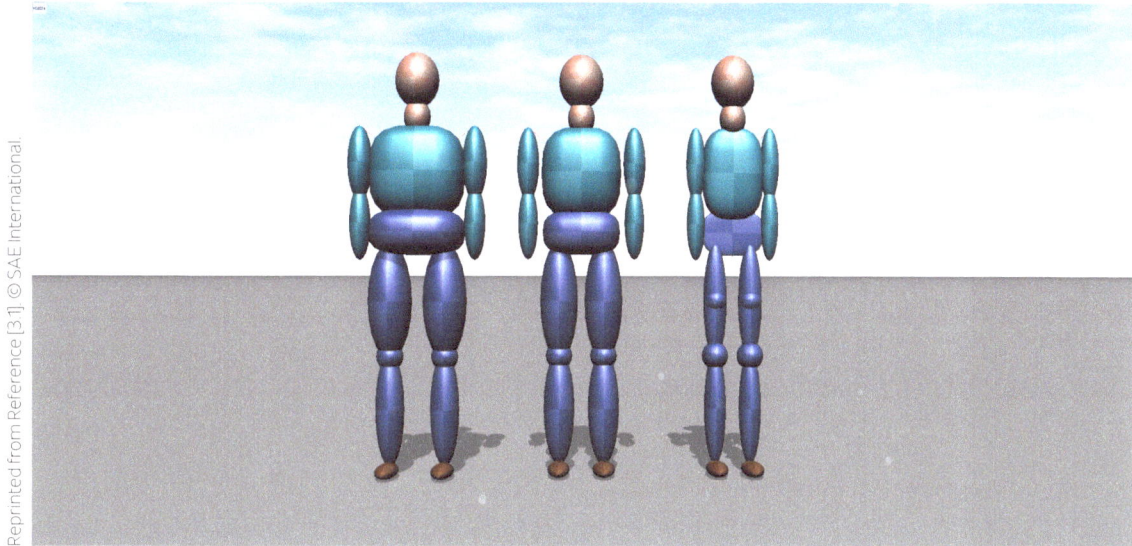

Alternatively, **Figure 3.3** shows an example of three pedestrian multibodies, again with weights of 250 lb (113.4 kg), 176.4 lb (80 kg), and 100 lb (45.4 kg), but this time these weights were set in the System Properties dialog box (contrary to what is recommended). When changed in this way, PC-Crash did not scale the size of the ellipsoids. This confirms the recommendation in the manual that the pedestrian model should be scaled using the Body Data dialog box. Note also that when the user resizes the pedestrian in the Body Data dialog box, PC-Crash does not automatically ensure that the pedestrian is positioned along the vertical axis to be on the terrain. The pedestrian can end up floating above the terrain or partially below the terrain. This influences the results of the simulation, so after resizing a pedestrian, ensure that the pedestrian's feet are positioned on the terrain.

Figure 3.3 Pedestrian parameters changed in the System Properties dialog box.

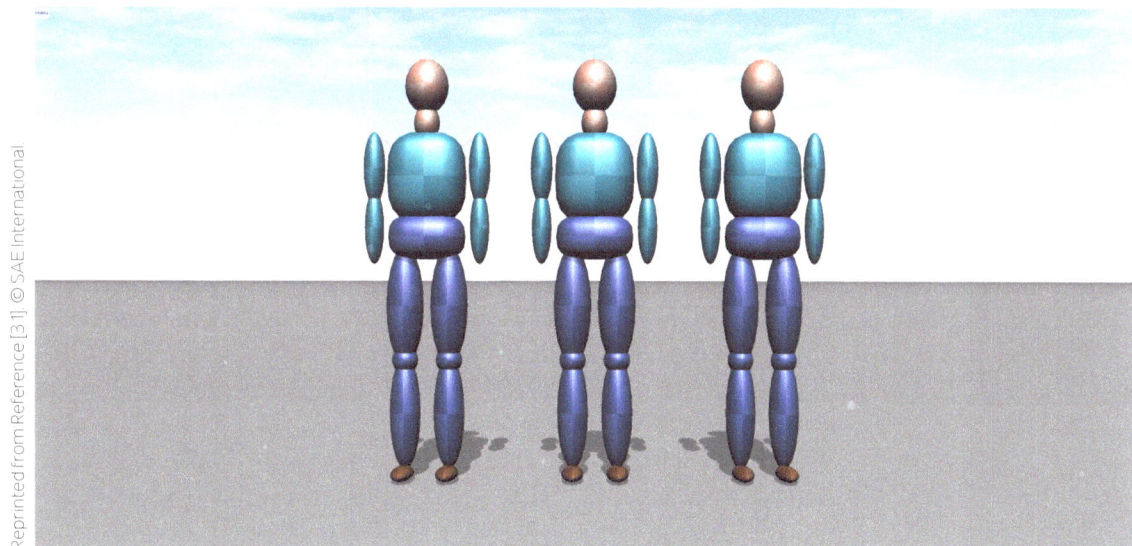

The pedestrian model consists of 20 bodies and 19 joints. The geometric and inertial parameters of each of the bodies can be modified by the user, though this would not usually be necessary. With its default sizing, the pedestrian multibody has a CG height of 39.9 in. (1.0 m) above the ground. The multibody has a default restitution value of 0.316, a default friction coefficient for interaction with the ground of 0.600, and a default friction coefficient for interaction with vehicles of 0.200. The CG height of the model was analyzed when it was sized to represent 1st, 50th, and 99th percentile males and females. The height and

weight for these percentiles were entered from a reference set of anthropometric data, and the resulting CG height from the scaled PC-Crash multibody pedestrian was determined. The resulting CG height for each configuration was compared to the CG heights reported in the anthropometric data [3.8]. **Tables 3.1** and **3.2** show the resulting comparison. This process revealed that the CG height in PC-Crash was the same for ages 20 to 60, but different for an entered age of 18. Overall, the CG heights of the PC-Crash pedestrian model showed adequate agreement with the referenced anthropometric data.

Table 3.1 CG calculation results, male.

	Anthropometric data			PC-Crash multibody
	Height (in.)	Weight (lb)	CG height (in.)	CG height (in.)
Male 18 years old 1st percentile	62.6	100.3	34.2	33.7
Male 20–60 years old 1st percentile	62.6	100.3		35.5
Male 18 years old 50th percentile	69.1	172.0	37.9	37.2
Male 20–60 years old 50th percentile	69.1	172.0		37.2
Male 18 years old 99th percentile	75.6	244.0	41.8	40.4
Male 20–60 years old 99th percentile	75.6	244.0		39.8

© SAE International

Table 3.2 CG calculation results, female.

	Anthropometric data			PC-Crash multibody
	Height (in.)	Weight (lb)	CG height (in.)	CG height (in.)
Female 18 years old 1st percentile	58.1	93.0	31.2	31.3
Female 20–60 years old 1st percentile	58.1	93.0		33.2
Female 18 years old 50th percentile	64.0	137.5	35.2	33.9
Female 20–60 years old 50th percentile	64.0	137.5		35.0
Female 18 years old 99th percentile	69.8	217.6	38.6	36.8
Female 20–60 years old 99th percentile	69.8	217.6		36.9

© SAE International

Simulations of a Pedestrian Sliding to Rest

This section describes simulations run with the PC-Crash multibody pedestrian model without any interaction with a vehicle. These simulations provide guidance on the input for the pedestrian-to-ground friction in PC-Crash. The reader is encouraged to refer to the related discussion in Chapter 1. In the simulations described in this section, the pedestrian was initially upright and facing up on the screen (viewed from a top-down perspective). The pedestrian was initially assigned a ground friction coefficient of 0.600 and a coefficient of restitution of 0.316 (the default values). The pedestrian was assigned an initial lateral velocity (left to right on the screen) and was allowed to fall, tumble, and slide to rest. An integration timestep of 0.5 ms was used. A baseline simulation was run with an initial speed of 45 mph (72.4 km/h). The pedestrian motion for this simulation is depicted in **Figure 3.4** at 0.25-sec intervals. The resultant torso and hip velocities of the pedestrian in this baseline simulation are depicted in **Figure 3.5**. The duration of the simulated pedestrian motion was 3.43 sec, and the pedestrian traveled 117 ft. Approximately 1.55 sec into the simulation, the pedestrian motion had stabilized to a pure slide

(vertical velocities went to zero at this time). The deceleration of the pedestrian from this point forward in the simulation was 0.6g, consistent with the ground friction input for the multibody model of 0.600. The pedestrian impacted the ground at approximately 0.43 sec into the simulation. At this time, the pedestrian model had a speed of approximately 44.43 mph based on the torso and hip velocities. Thus, including the ground impact in the deceleration yielded a deceleration of 0.67g.

The overall scene friction input for this baseline simulation was the default value of 0.8. This was changed to 1.0, and the simulation was run again. No changes occurred in the simulation. Next, the overall scene friction was changed to 0.1, and the simulation was run again. No changes occurred in the simulation. These simulations confirmed that scene friction input does not control or influence pedestrian deceleration. Similarly, a friction zone was created in PC-Crash that covered the entire area of the ground where pedestrian motion occurred. The friction of this zone was set to 2.0, and the simulation was rerun. Again, no changes occurred in the simulation, confirming that friction zones also do not influence pedestrian deceleration.

Figure 3.4 Baseline simulation motion.

Figure 3.5 Baseline simulation, torso and hip resultant velocities.

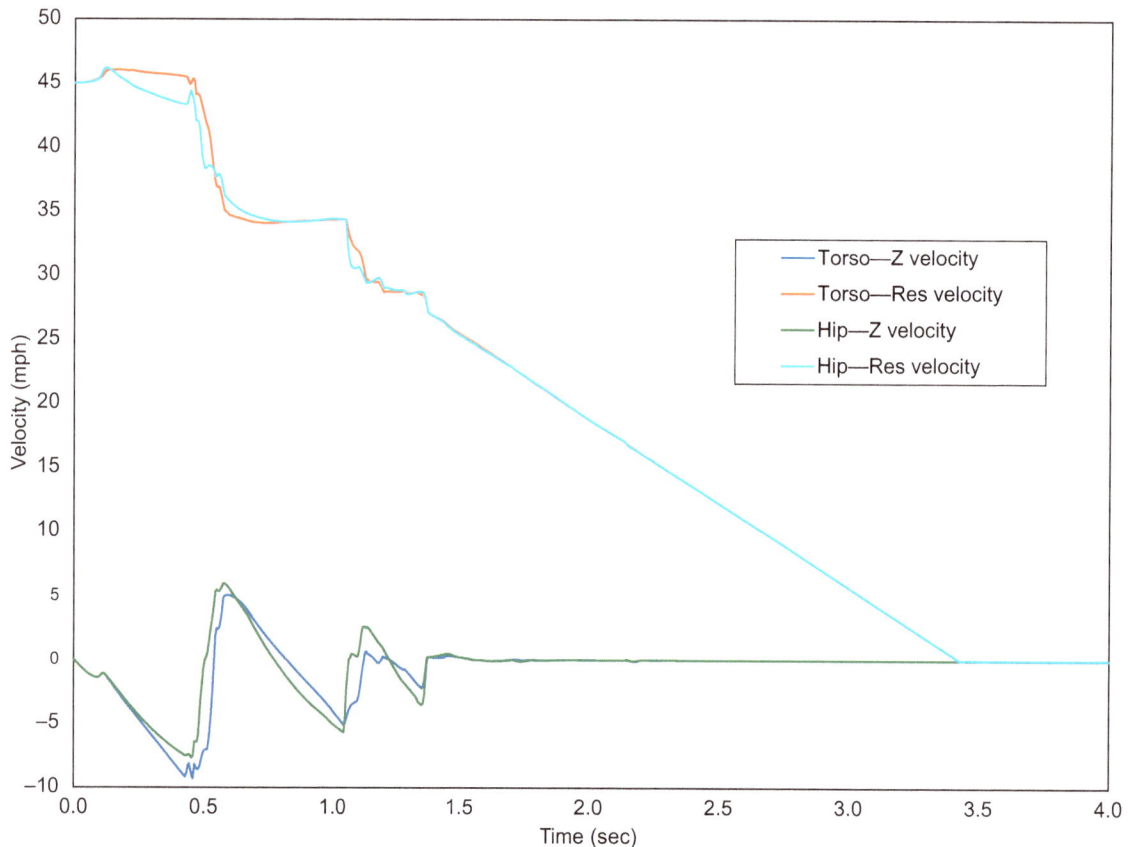

Next, the initial lateral speed of the pedestrian in the baseline simulation was changed to 25 mph (40.2 km/h). All other variables remained the same. An integration timestep of 0.5 ms was used, and the duration of the simulated pedestrian motion was 1.98 sec. Approximately 1.8 sec into the simulation, pedestrian motion had stabilized to a pure slide. The deceleration of the pedestrian from this point forward in the simulation was approximately 0.6g, consistent with the ground friction input of 0.600. The pedestrian impacted the ground at approximately 0.43 sec into the simulation. At this time, the pedestrian had a speed of approximately 25 mph (40.2 km/h) based on the torso and hip

velocities. Thus, including the ground impact in the deceleration yielded a deceleration of 0.74g. This is a higher value than the prior simulation, consistent with the fact that, at lower speeds, the ground impact makes a greater contribution to the overall deceleration.

An additional simulation was run from an initial pedestrian speed of 45 mph. For this simulation, the friction coefficient between the pedestrian and the ground was lowered to 0.40 in the baseline simulation. All other parameters were held fixed. The duration of the simulated pedestrian motion was 5.13 sec, and the pedestrian traveled 173 ft. Approximately 1.5 sec into the simulation, pedestrian motion had stabilized to a pure slide. The deceleration of the pedestrian from this point forward in the simulation was 0.4g, consistent with the ground friction input for the multibody model of 0.400. The pedestrian impacted the ground at approximately 0.46 sec into the simulation. At this time, the pedestrian model had a speed of approximately 44.56 mph based on the torso and hip velocities. Thus, including the ground impact in the deceleration yielded a deceleration of 0.434g.

These simulations confirmed that the pedestrian-to-ground friction coefficient for the PC-Crash pedestrian model should be set using values representative of simple sliding, not values that include the impact with the ground or the airborne trajectory. The airborne and landing phases are accounted for in the physics models of PC-Crash, and they do not need to be accounted for in the ground friction input to the pedestrian model. This, unfortunately, invalidates the numerical results from Richardson et al. [3.6]. These authors set the ground friction for the pedestrian to 0.8 or 1.2, values that would include the deceleration from the pedestrian's impact with the ground. Entering such high values for the ground friction amounts to double-counting the speed loss from the ground impact. The underlying finding of the Richardson et al. study—that there is an influence of the vehicle shape and impact location on the pedestrian throw distance and post-impact travel distance—is valid, but the numerical values in the study should not be used.

Simulating Forward Projection Collisions in PC-Crash

Now, consider the testing conducted by Fugger et al. [3.9]. These authors reported 140 pedestrian impact crash tests utilizing high-fronted vehicles that generated forward projection trajectories. An Alderson Research Labs CG-95 dummy was utilized, and the dummy was clothed in a wetsuit covered by coveralls and standard athletic footwear. This dummy was 75.5 in. (1.92 m) tall and weighed 169 lb (76.7 kg). Of the 140 tests, 56 were conducted on dry asphalt and 84 on wet asphalt. Impact speeds varied between 4 km/h (2.5 mph) and 60 km/h (37 mph), with most of the tests conducted at speeds below 32 km/h (20 mph). **Figure 3.6** presents a graph depicting the throw distance results from Fugger et al.'s testing. Throw distance in meters is plotted on the horizontal axis, and impact speed in kilometers per hour is plotted on the vertical axis. The different colors of circles represent tests run with varying surface conditions. The tests run on the dry surface are shown with orange circles, and those run on wet surfaces are shown with gray, yellow, and blue circles. The gray circles are for a wet surface with less than 1/16th of an inch of water, and the yellow circles for a surface with more than 1/16th of an inch of water. No description was given in the study for the surface called "Wet II" (blue circles).

Figure 3.6 Throw distance versus speed data from the testing by Fugger et al.

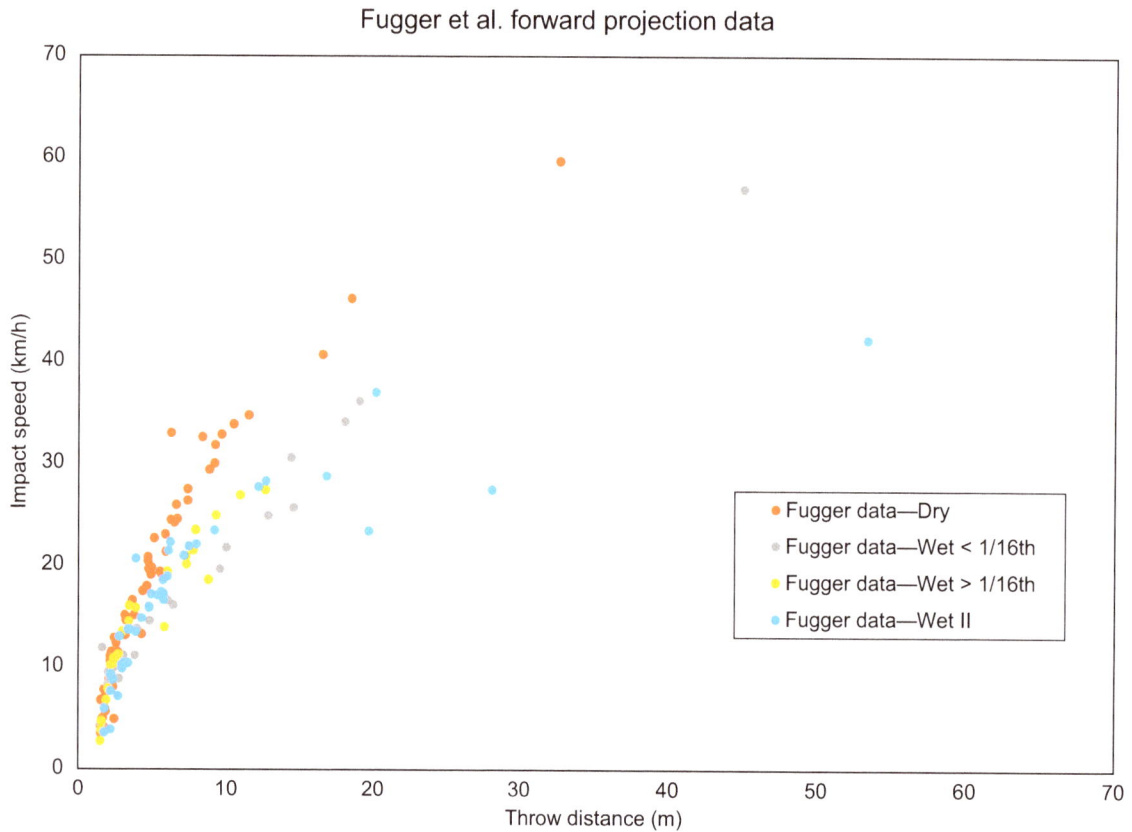

Fugger et al. forward projection data

Fugger et al. reported that the tests run on dry asphalt utilized either a 1976 Ford Econoline 250 van or a 1980 Plymouth D100 van. The dimensions and weights of these vehicles were not reported in the study, but vehicle identification numbers (VINs) were. Based on manufacturer specifications, the curb weights of these vehicles were estimated at 4706 and 4140 lb (2134.6 kg and 1877.9 kg), respectively. These curb weights are listed in **Table 3.3** along with the overall dimensions of these vehicles. The tests run on wet asphalt utilized a 1971 Dodge B200 van, a 1982 Dodge B250 van, a 1982 Chevy G20 van, and a 1974 Dodge Sportsman van. The curb weights of these vehicles were estimated at 4025, 3730, 3830, and 4025 lb (1825.7, 1691.9, 1737.3, and 1825.7 kg), respectively. These curb weights are listed in **Table 3.3** along with the overall dimensions of these vehicles. Fugger et al. did not report which individual tests were run with which vehicle. However, the entire range of test vehicle weights significantly exceeded the weight of the pedestrian dummy, and so the simulations were not expected to exhibit sensitivity to the specific vehicle. In addition, the widths and heights of the six vehicles were similar. For the initial set of simulations using the Fugger et al. test protocol, a representative vehicle weight of 4000 lb was utilized. Similarly,

representative dimensions of 205.9, 79.1, and 78.3 in. were utilized for the overall length, width, and height. These dimensions produced forward projection trajectories in the simulations, consistent with the description of the pedestrian trajectories given in the Fugger et al. study.

Table 3.3 Test vehicles utilized by Fugger et al.

	Surface condition	Curb weight (lb)	Length (in.)	Width (in.)	Height (in.)
1976 Ford Econoline E250	Dry	4706	228	80	82
1980 Plymouth Voyager	Dry	4140	197	80	81
1971 Dodge B200	Wet <1/16th	4025	197	80	81
1982 Dodge B250 LWB	Wet >1/16th	3730	196.9	79.8	80.9
1982 Chevrolet G20	Wet II	3830	202.2	79.5	79
1974 Dodge Sportsman	Wet II	4025	194	79.8	81

Fugger et al. reported that the upper leading edge of the test vehicles was situated above the CG of the pedestrian dummy. The graphic presented in their study is included in **Figure 3.7**, showing the front profiles of five of the six vans utilized in the study. As this graphic shows, there was variability in the distance from the front of the vehicle to the base of the windshield. However, the heights of the upper leading edge of the

vehicles were similar, and so in terms of the throw distances generated, PC-Crash simulations were not expected to exhibit sensitivity to the specific front profile utilized, as long as the profile was similar and had an upper leading edge well above the CG of the pedestrian. A DXF model of a Ford E-150 was used for these simulations (**Figure 3.8**).

Figure 3.7 Van front profiles from the Fugger et al. study.

The pedestrian multibody was sized to the height and weight reported for the dummy in the Fugger et al. study. Simulations were then run with pedestrian-to-ground friction coefficients of 0.3, 0.4, 0.6 (the default value), and 0.65 with impact speeds of 5, 7.5, 10, 15, 20, 25, 30, 35, and 40 mph (8.0, 12.1, 16.1, 24.1, 32.2, 40.2, 48.3, 56.3, and 64.4 km/h). The simulations were run with the default pedestrian-to-vehicle friction coefficient of 0.2 and, initially, the default restitution value of 0.316. The restitution input was later varied. An integration time step of 0.1 ms was used. Heavy braking was assumed to be present for the van during and following each collision. **Figure 3.8** depicts a series of frames from one of these simulations (van impact speed = 40 mph, pedestrian-to-ground friction = 0.65). The first frame in this figure is at the start of the simulation; the second frame is 60 ms into the simulation when the pedestrian is fully engaged with the front of the vehicle; the third frame is 390 ms into the simulation when the pedestrian has separated from the vehicle and is airborne; and the fourth frame is 975 ms into the simulation when the pedestrian has first contacted the ground.

Figure 3.8 Graphics showing simulation motion for a forward projection simulation.

(a)

(b)

(c)

(d)

Figure 3.9 shows the results of simulations for pedestrian-to-ground friction coefficients of 0.65 and 0.6, along with the default values for car-to-pedestrian friction and restitution. This graph also includes the tests from the Fugger et al. study for the dry surface. Throw distance in meters is plotted on the horizontal axis, and impact speed in kilometers per hour is plotted on the vertical axis. Circular points on the graph are tests reported in the Fugger et al. study, and square points are PC-Crash simulations. This graph also includes the 15th and 85th percentile bounds for the forward projection equation reported by Toor and Araszewski [3.10], which is the following:

$$V_v = 8.25S^{0.61} \pm 7.7 \text{ km / h.}$$

(3.1)

Figure 3.9 Throw distances from the Fugger et al. study and PC-Crash, rest. = 0.316 (dry).

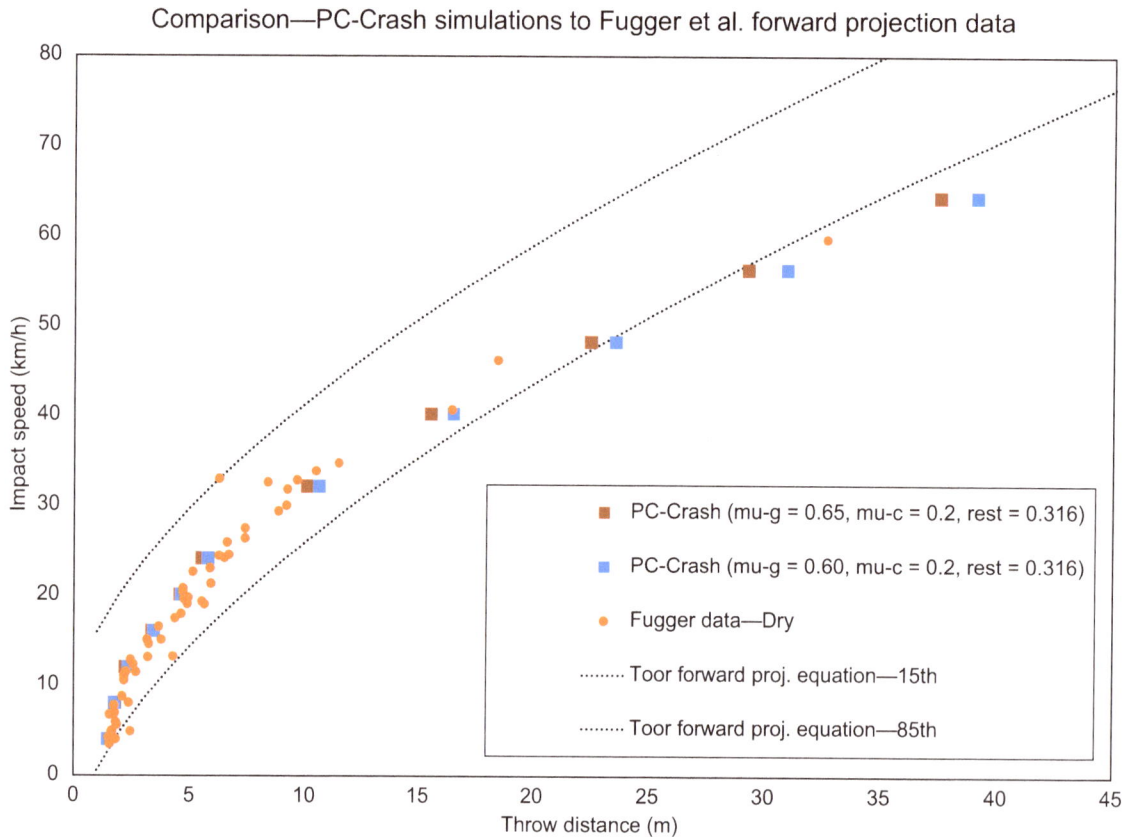

Examination of **Figure 3.9** reveals that, for speeds under approximately 25 mph (40 km/h), there was agreement between the trend in the Fugger et al. data and the trend in the simulations. For speeds more than approximately 25 mph (40 km/h), the pedestrian in the PC-Crash simulations generally traveled farther than the dummies in the Fugger et al. dataset and farther than what Equation (3.1) would predict. These simulations were all run with PC-Crash's default restitution value of 0.316. Normally, restitution

decreases with increasing collision severity, and so, while the PC-Crash default of 0.316 is appropriate for lower-speed collisions, based on the Fugger et al. data, it is too high for higher-speed collisions. **Figure 3.10** again shows the results of PC-Crash simulations for pedestrian-to-ground friction coefficients of 0.65 and 0.6, along with the tests from the Fugger et al. study for the dry surface. This time, though, the simulations were run with a coefficient of restitution for the pedestrian of 0.000 or 0.100.

Figure 3.10 Throw distances from the Fugger et al. study and PC-Crash, reduced restitution (dry).

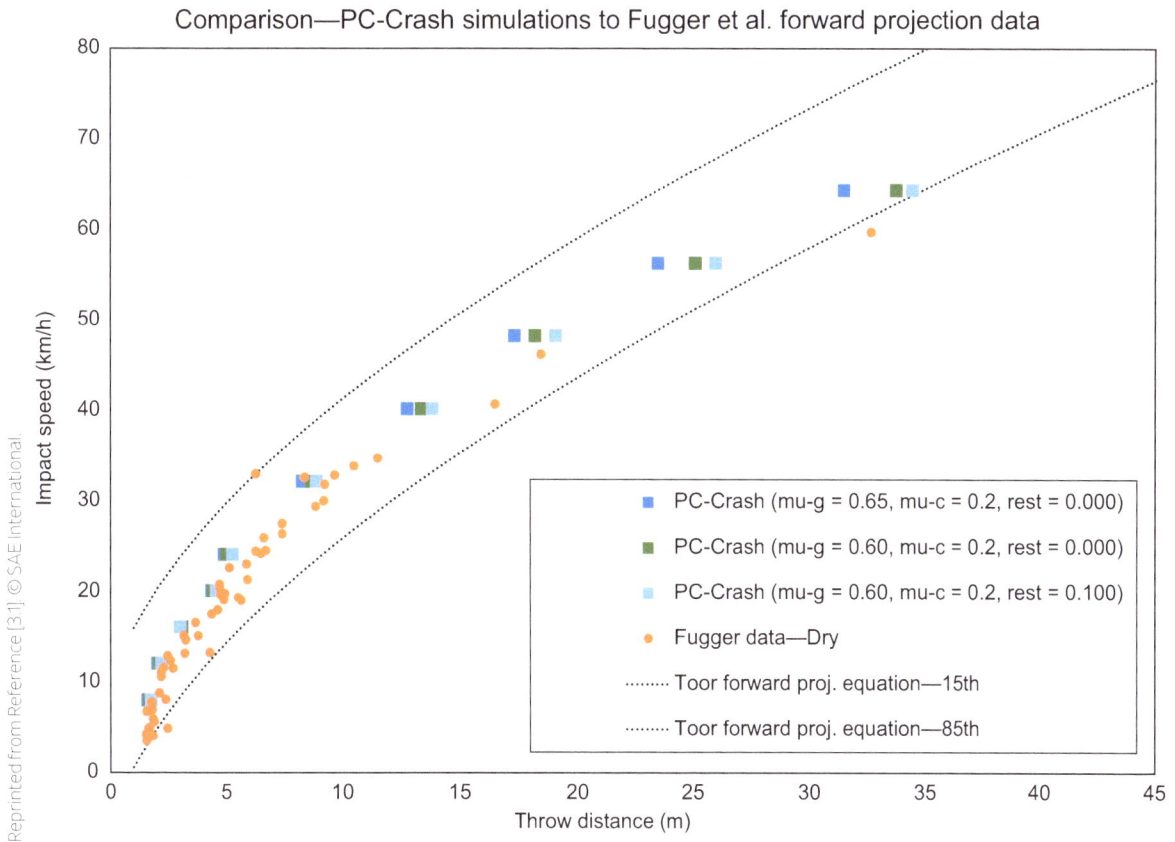

This time, the throw distances predicted by the simulations fell within the 15th and 85th percentile bounds of Equation (3.1), and the trend of the simulations with a coefficient of restitution of 0.100 showed agreement with the trend of the higher-speed Fugger et al. data. However, the agreement with the test data was degraded at lower speeds with lower coefficients of restitution. This result demonstrates that the default restitution input of 0.316 for the PC-Crash pedestrian model is appropriate for lower-speed collisions but needs to be lowered for higher-speed forward projection collisions. In this instance, the cutoff between the "lower" and "higher" speed collisions was approximately 25 mph (40 km/h).

Figure 3.11 shows the Fugger et al. data for the wet surfaces plotted with the 15th and 85th percentiles of the Toor and Araszewski forward projection equation. This comparison confirms that Equation (3.1) is not applicable to forward projection equations occurring on wet surfaces. **Figure 3.11** also includes the results of PC-Crash simulations for pedestrian-to-ground friction coefficients of 0.4 and 0.3. These simulations utilized the default pedestrian-to-vehicle friction coefficient of 0.2 and coefficient of restitution of 0.316. The trend in the throw distances predicted by these simulations showed agreement with the trend in the throw distances from the Fugger et al. dataset. In this instance, there are limited data points beyond 30 km/h and considerable variability in the data in this speed range.

Figure 3.11 Throw distances from the Fugger et al. study and PC-Crash (wet).

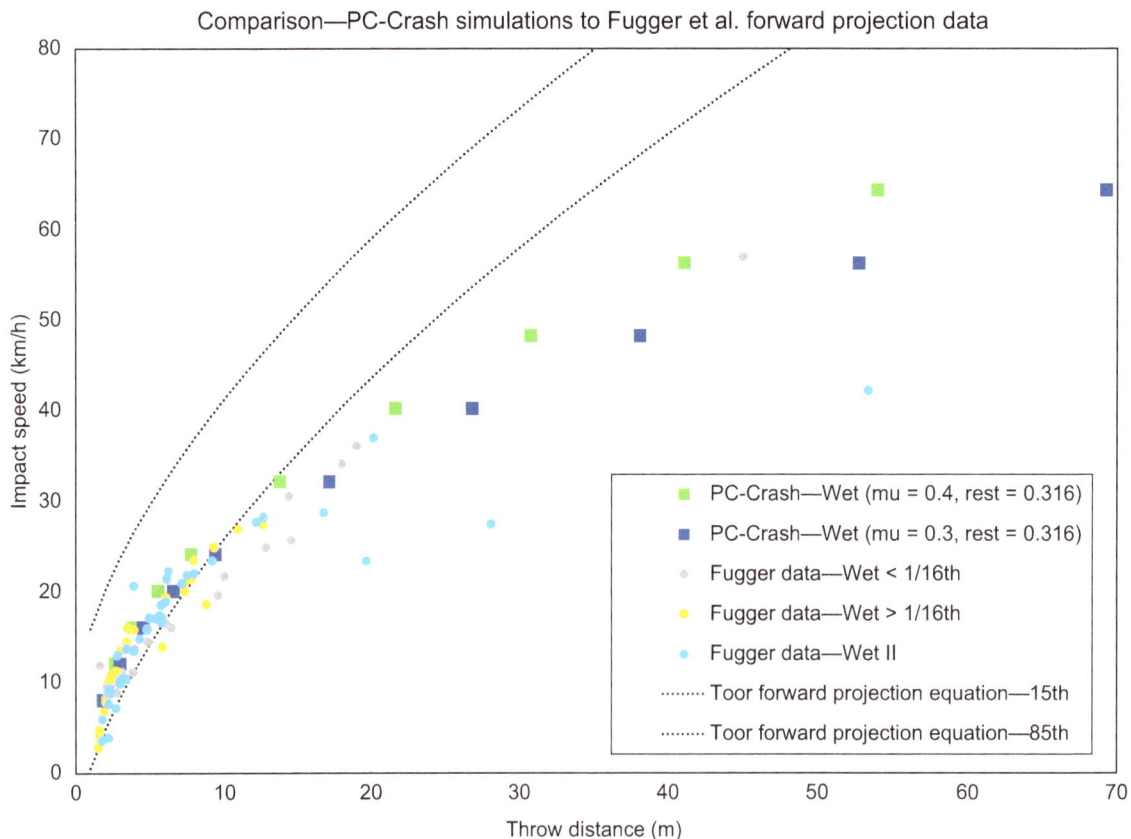

Simulating Wrap Trajectories in PC-Crash

Now, consider the testing reported by Severy and Brink [3.11]. These authors reported staged pedestrian collisions in which 38 anthropometric dummies of varying size (adult, six-year-old, three-year-old, and toddler) and posture were impacted by vehicles of varying size and shape (sports car, passenger car, and truck). Impacts speeds of 10, 20, 30, and 40 mph (16.1, 32.2, 48.3, and 64.4 km/h) were utilized. The position of the dummies relative to the front of the vehicle was varied (glancing, fender, off-center, and center). Braking levels at impact were varied (none, moderate, or panic braking). A friction coefficient for the test surface was not reported in the study, nor were the maximum decelerations of the vehicles. The vehicle speed for each test was reported with the nominal value, but the brakes were reportedly applied just prior to impact. There was likely some speed loss prior to impact, and therefore, the actual impact speeds were likely at or below the reported nominal values. For the PC-Crash simulations, a coefficient of friction between the vehicle tires and the test surface of 0.8 was utilized. The speed of the test vehicle at the impact was initially assumed to be the nominal reported value, but in some of the simulations, small speed reductions were needed in order for the test vehicle to come to rest in the reported location. **Table 3.4** lists the weights and basic dimensions of the test vehicles. **Table 3.5** lists the weights and basic dimensions of the test dummies. For the simulations discussed here, only staged collisions with the adult dummy were simulated. The goal of these simulations was to determine the friction and restitution inputs for the pedestrian model that yielded the best match with the throw distance of the dummy, including both the longitudinal and lateral components.

Table 3.4 Specs of vehicles used in the Severy and Brink tests.

	1956 Corvette	1963 Plymouth	1963 Ford Truck
Weight (lb)	3400	3400	5000
Bumper height (in.)	19	17.5	25
Hood height (in.)	24	30	52
Bumper extension (in.)	4	1	6

© SAE International

Table 3.5 Specs of pedestrian models used in the Severy and Brink tests.

	Toddler manikin	Three-year-old manikin	Six-year-old manikin	Adult anthro dummy
Weight (lb)	32	35	48	200
Height (in.)	29	36	46	72
Head weight (lb)	4	4.5	6.5	12
CG height (in.)	17	20	25	41

© SAE International

Test 65

Test 65 was simulated in PC-Crash with an integration timestep of 0.1 ms. This test involved the 1963 Plymouth traveling at approximately 20 mph (32 km/h) at the impact with the brakes applied. The vehicle in this test struck four dummies: an adult dummy at the right front, two three-year-old dummies, and a toddler dummy. Only the adult dummy was simulated. The PC-Crash pedestrian was sized according to the height and weight listed in **Table 3.5**. The vehicle was set up in accordance with the specifications listed in **Table 3.4**. A DXF model available in PC-Crash for a 1968 Barracuda Formula S 383 was used. **Figure 1.8** presents a diagram for Test 65, which shows the impact and rest positions of the Plymouth and the dummies. This scaled diagram was imported into PC-Crash and used for setting up and optimizing the simulation of this test. In setting up the simulation, the default friction and restitution inputs were used initially. Based on the findings of the forward projection collisions, the impact speed for this test was still within the range where the default coefficient of restitution of 0.316 was plausible. We experimented with changing the pedestrian-to-vehicle coefficient of friction—increasing it from the 0.2 default value to 0.3 and 0.4. This friction coefficient had a negligible influence on the total throw distance of the pedestrian. Of the three friction/restitution inputs, the pedestrian-to-ground friction was the most influential in determining the pedestrian's total throw distance in the simulation, and a value of 0.50 for this parameter gave the best match with the throw distance.

The images in **Figure 3.12** are two video frames from the study that informed the setup of the simulation of this test. The images in the Severy and Brink study were limited, and so there was some uncertainty in setting up the pedestrian multibody at impact. It was apparent from the available images, though, that the pedestrian dummy was positioned at the passenger's side front corner of the test vehicle facing toward the driver's side with its left leg forward and its rear leg back. The left side of the dummy was struck by the test vehicle. The precise angles of the limbs were not known. In optimizing the simulation, small adjustments to the precise location of the dummy at impact were made. Moving the multibody laterally toward the passenger's side of the vehicle increased the lateral movement of the pedestrian from the vehicle, so the precise lateral position was set to optimize the actual lateral movement from the test. The needed adjustments were small, and the final optimized impact position of the dummy was consistent with the position depicted in **Figure 1.8**.

Figure 3.12 Video frames from Test 65.

(a)

(b)

In the final optimized simulation, both the longitudinal and lateral throw distances of the dummy were matched, but the specific orientation of the dummy was not. A match was sought with the initial motion of the dummy relative to the vehicle and with the total throw distance. Once the dummy separated from the vehicle, the specific orientation of the dummy and the individual limbs was not used to judge the quality of the simulation. The images in **Figure 3.13** show the setup of the vehicle and

pedestrian in PC-Crash for the start of the simulation. **Figure 3.14** shows the resulting rest positions of the vehicle and the pedestrian. During the simulation process, the post-collision brake factors for the test vehicle were set to achieve a match with the actual rest position of the vehicle. In this instance, a post-impact deceleration of 0.4g was generated with the brake factors.

Figure 3.13 Test 65 simulation setup.

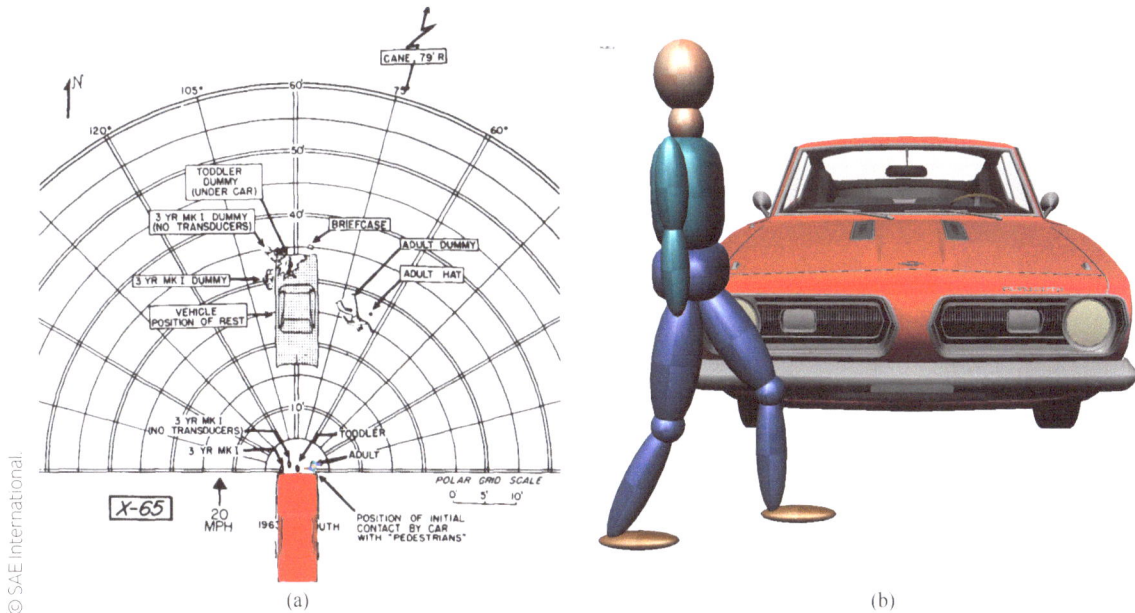

(a)

(b)

Figure 3.14 Rest positions from optimized simulation (Test 65).

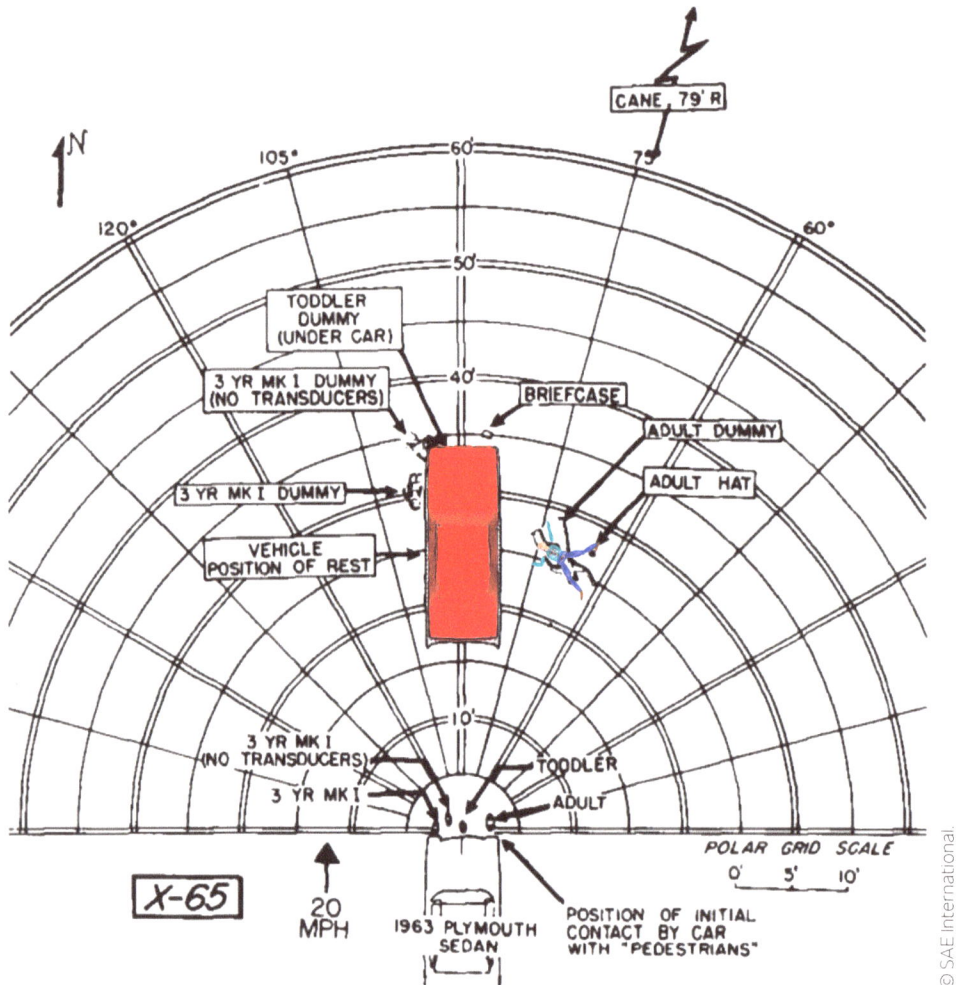

© SAE International

Test 67

A similar procedure was followed in simulating additional tests from the Severy and Brink study, including Test 67 (**Figure 1.1**). This test involved the 1963 Plymouth traveling at 30 mph (48 km/h) at the impact with the brakes applied. The vehicle in this test struck four dummies: an adult dummy at the right front, two toddler dummies, and a three-year-old dummy. Only the adult dummy was simulated. The PC-Crash multibody

pedestrian was again sized according to the height and weight listed in **Table 3.5**, and the vehicle was set up using the specifications listed in **Table 3.4**. A DXF model available in PC-Crash for a 1968 Barracuda Formula S 383 was again used. Video images from the Severy and Brink study were reviewed to assist with the setup of the dummy's initial posture and position. In this instance, the best match was obtained with a pedestrian-to-ground friction coefficient of 0.65, a pedestrian-to-vehicle coefficient of friction of 0.2, and a coefficient of restitution of 0.04. **Figure 3.15** shows a top-down view of the resulting rest positions of the vehicle and the

pedestrian. In this instance, a post-impact deceleration of 0.67g was needed to match the actual rest position of the vehicle.

Figure 3.15 Rest positions from optimized simulation (Test 67).

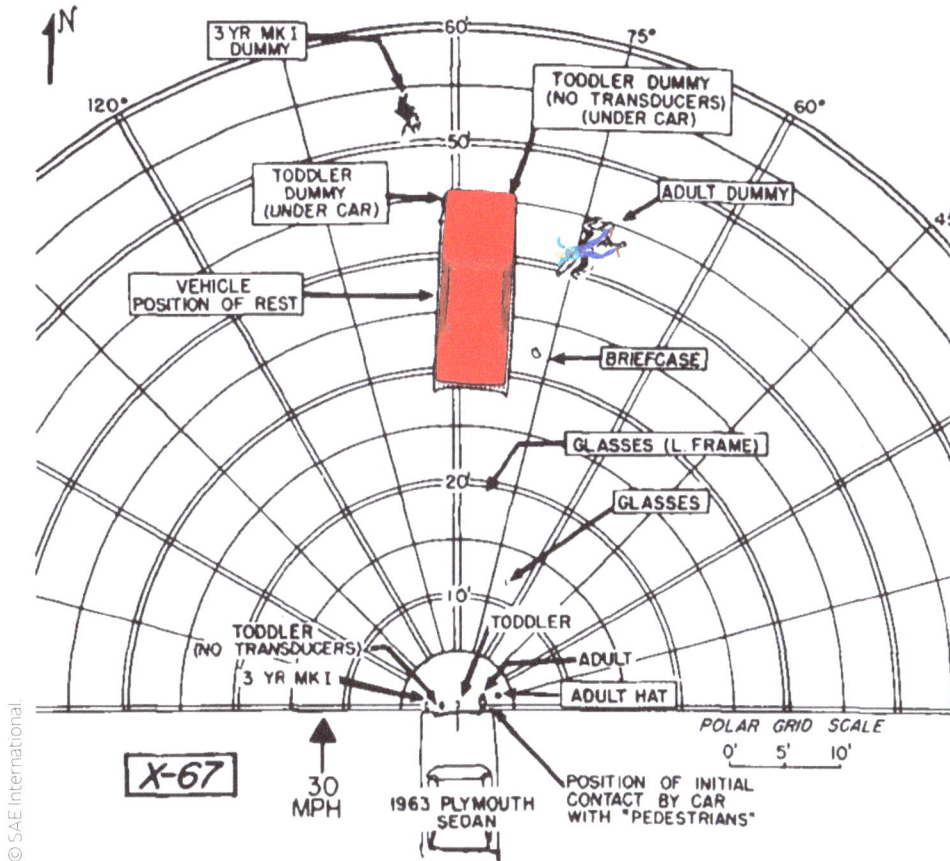

Test 71

Test 71 was simulated next. This test involved the 1963 Plymouth traveling at 30 mph (48 km/h) at the impact with the brakes applied. The vehicle struck four dummies, but only the adult dummy was simulated. The multibody pedestrian was again sized according to the height and weight listed in **Table 3.5**, and the vehicle was set up using the specifications listed in **Table 3.4**. The DXF model in PC-Crash for a 1968 Barracuda

Formula S 383 was again used. In this instance, the best match was obtained with a pedestrian-to-ground friction coefficient of 0.65, a pedestrian-to-vehicle coefficient of friction of 0.2, and a coefficient of restitution of 0.175. **Figure 3.16** shows the resulting rest positions of the vehicle and the pedestrian. A post-impact deceleration of 0.79g was necessary to match the rest position of the vehicle. Uneven left-to-right braking was also needed.

Figure 3.16 Rest positions from optimized simulation (Test 71).

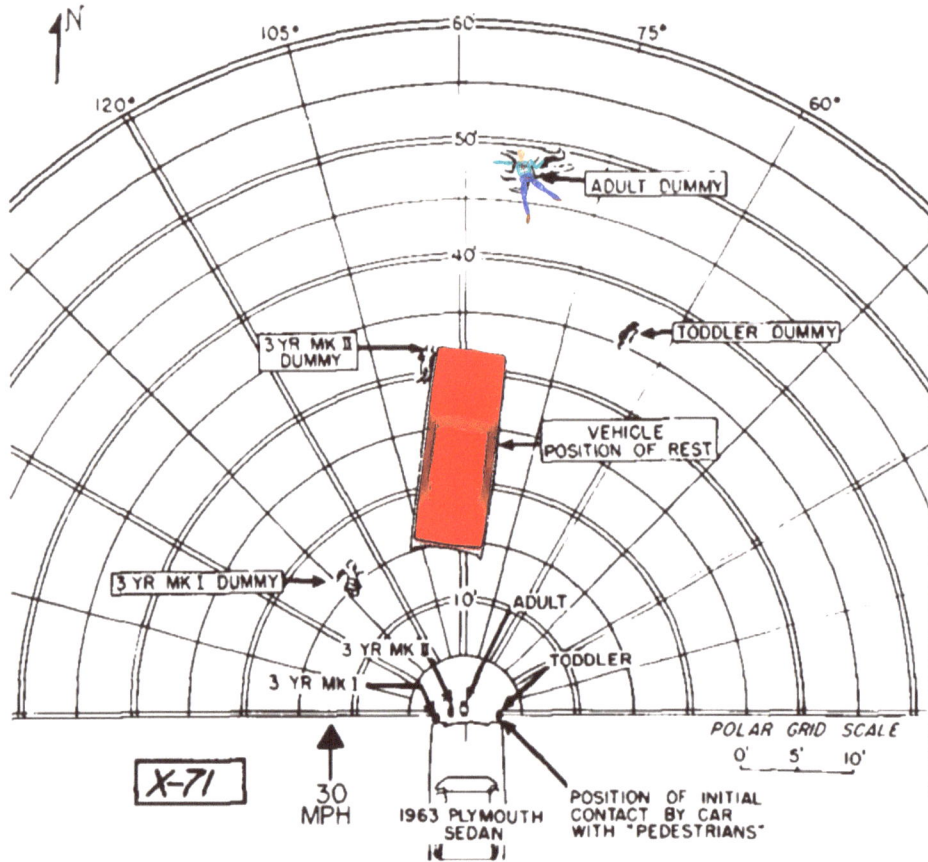

© SAE International.

Simulation of Additional Staged Collisions

The authors of this book ran a series of staged pedestrian collisions, two of which were previously discussed in Chapters 1 and 2. These tests involved a 2007 Chevrolet Malibu impacting a full-sized pedestrian dummy. **Figure 1.2** depicts the test setup. Given the front profile of the vehicle and the size of the dummy, these tests generated wrap or fender vault trajectories for the dummy. After each test, the vehicle rest position, dummy rest position, and debris locations were documented with terrestrial photographs and aerial photographs.

GPS-enabled Propeller AeroPoints[1] were placed along the test area to provide ground control, and aerial photographs were taken in a manner conducive to aerial mapping [3.12]. The vehicle and pedestrian at rest were also scanned with an iPhone 13 Pro and Recon3D software.[2]

Test 2

In Test 2, the dummy was placed in the center of the test fixture to align with the center of the test vehicle. A canvas shopping bag was affixed to the dummy's left hand and loaded with bottles of water. The driver accelerated the vehicle to a speed of 49 mph and then applied the brakes aggressively

---

1   https://www.propelleraero.com/aeropoints/

2   https://www.recon-3d.com/

prior to the contact with the dummy. The impact speed of the vehicle was approximately 40.1 mph. The dummy contacted the front bumper and grille of the vehicle and then wrapped onto the hood and windshield in a location near the longitudinal centerline of the vehicle with a slight offset to the passenger side. **Figure 3.17** is a video frame that

depicts the first contact between the dummy and vehicle. **Figure 1.3** is another video frame depicting the dummy being wrapped onto the front of the vehicle and in contact with the windshield. Along the longitudinal axis of the vehicle, the dummy's head struck the vehicle at approximately the center of the windshield.

Figure 3.17 Impact configuration for Test 2.

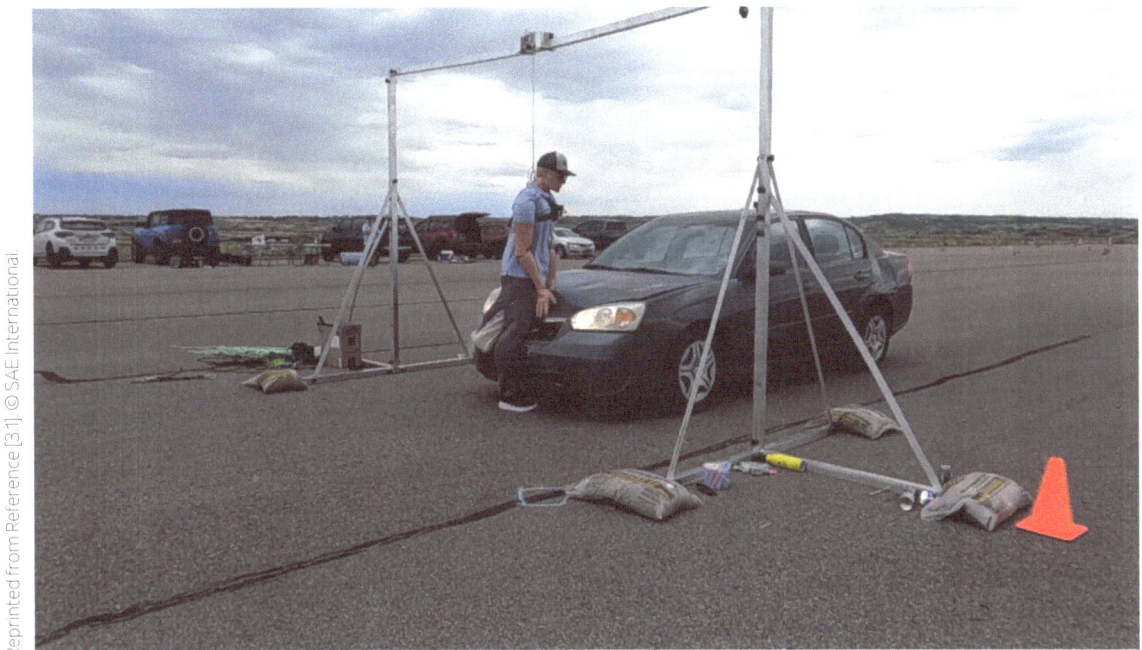

Reprinted from Reference [31]. © SAE International.

From first contact to rest, the Malibu in this test traveled approximately 57 ft and the test dummy traveled approximately 81 ft. The bag was impacted by the vehicle and was projected forward from the impact point and was located near the point of rest of the vehicle. The dummy's hat came to rest near the point of impact. The diagram in **Figure 1.5** is an orthomosaic created using Pix4D mapper from the aerial imagery taken after the test. Labels have been added to

identify the vehicle rest position, pedestrian rest position, debris locations, and tire marks. This test was simulated with the PC-Crash multibody pedestrian model with an integration timestep of 0.1 ms. The PC-Crash multibody pedestrian was scaled to the height and weight of the dummy used in the tests. A stock Chevrolet Malibu DXF model available in PC-Crash was utilized. **Figure 3.18** shows the setup of the vehicle and pedestrian in PC-Crash for the start of the simulation.

Figure 3.18 Simulation setup for Test 2.

In this instance, the best match with the longitudinal and lateral components of the throw distance and the contact points of the dummy on the vehicle was obtained with a pedestrian-to-ground friction coefficient of 0.6, a pedestrian-to-vehicle coefficient of friction of 0.2, and a coefficient of restitution of 0.13.

Figure 3.19 shows the resulting rest positions of the vehicle and the pedestrian. **Figure 3.20** shows where the dummy's head struck the vehicle in the simulation. During the simulation process, the post-collision brake factors for the test vehicle were set to achieve a match with the actual rest position of the vehicle.

Figure 3.19 Rest positions from optimized simulation (Test 2).

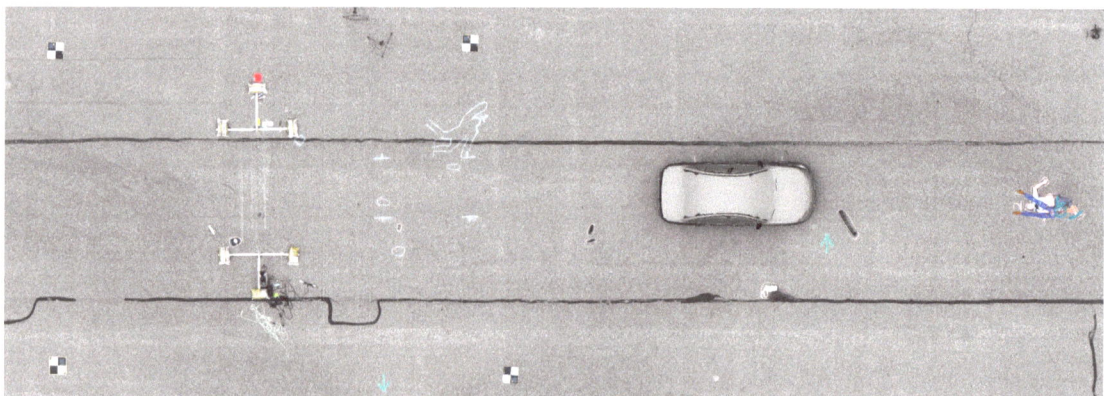

Figure 3.20 Dummy head contact location from the simulation (Test 2).

Test 3

In Test 3, the dummy was offset to the side of the test fixture to align with the passenger side corner of the test vehicle. A canvas shopping bag was affixed to the dummy's left hand and loaded with bottles of water. The driver accelerated the vehicle to a speed of 52 mph and then applied the brakes aggressively prior to the contact with the dummy. At the first contact, the vehicle was traveling at approximately 42.3 mph. The dummy contacted the front passenger side bumper of the vehicle and then wrapped onto the passenger side fender. The dummy's head contacted the A-pillar and windshield, before contacting the roof of the vehicle. **Figure 3.21** is a video frame that depicts the first contact between the dummy and vehicle. **Figure 1.4** is another video frame showing the dummy being wrapped onto the front of the vehicle. After the impact, the dummy was projected longitudinally and laterally away from the passenger side of the vehicle.

The diagram in **Figure 1.6** is an orthomosaic created from the aerial imagery taken after the test, with labels added to identify the vehicle rest position, pedestrian rest position, and debris locations. From the first contact to rest, the Malibu traveled approximately 69 ft and the test dummy traveled approximately 64 ft along the longitudinal axis of the vehicle and approximately 24 ft laterally to the vehicle for a total distance of 68 ft. The bag was not impacted by the vehicle and came to rest near the point of impact, as did the dummy's hat and sunglasses. This test was simulated in PC-Crash with an integration timestep of 0.1 ms. The multibody pedestrian was scaled to the height and weight of the dummy used in the tests. A stock Chevrolet Malibu DXF model available in PC-Crash was utilized. **Figure 3.22** shows the setup of the vehicle and pedestrian in PC-Crash for the start of the simulation.

Figure 3.21 Impact configuration for Test 3.

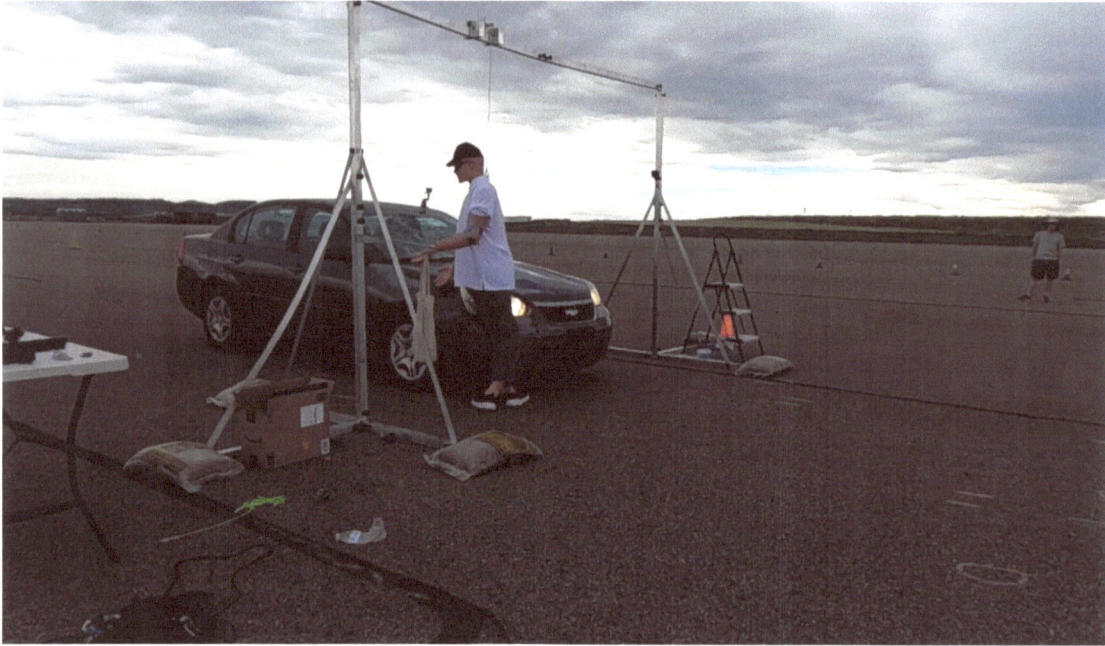

Figure 3.22 Simulation setup for Test 3.

In this instance, the best match was obtained with a pedestrian-to-ground friction coefficient of 0.5, a pedestrian-to-vehicle coefficient of friction of 0.1, and a coefficient of restitution of 0.316. **Figure 3.23** shows a top-down view of the resulting rest positions of the vehicle and the

pedestrian. **Figure 3.24** shows where the dummy's head struck the vehicle in the simulation. During the simulation process, the post-collision brake factors for the test vehicle were set to achieve a match with the actual rest position of the vehicle.

Figure 3.23 Rest positions from optimized simulation (Test 3).

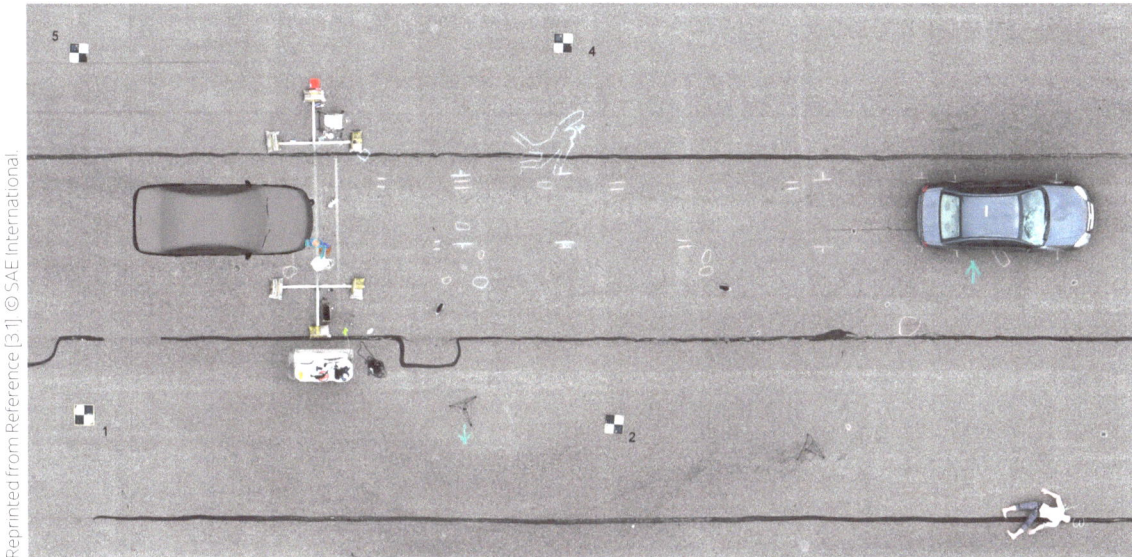

Figure 3.24 Dummy head contact location from the simulation (Test 3).

In closing this section, it is worth noting the simulation software Virtual Crash also includes a multibody pedestrian model. Becker, Reade, and Scurlock [3.13] validated the multibody model in this software for simulating pedestrian collisions. They reported that "the simulator faithfully reproduces the expected behavior of projectiles during both the airborne and ground sliding phases of their trajectories. The throw distances as a function of impact speed behavior of the simulated dummy model does a good job reproducing the behavior observed during staged impact experiments as well as that which was observed in prior experiments."

Conclusions

This chapter has described the multibody pedestrian model in PC-Crash, demonstrating that the model yields accurate simulations of vehicle–pedestrian collisions, particularly in terms of accurately simulating the contact points between the pedestrian and the vehicle and in predicting the total throw distance of the pedestrian. This following conclusions are presented:

1. The CG height of the PC-Crash pedestrian model was comparable to the CG heights reported for pedestrians in anthropometric data.

2. The pedestrian-to-ground friction coefficient should be set using values representative of simple sliding, not values that include the impact with the ground or the airborne trajectory.

3. Based on the simulations presented in this study, a reasonable range for this pedestrian-to-ground friction coefficient for dry roadways is 0.5 to 0.65.

4. For forward projection and wrap trajectories, the default coefficient of restitution for the multibody of 0.316 is reasonable for vehicle impact speeds below 40 km/h. For speeds

above 40 km/h with these trajectory types, a restitution coefficient in the range of 0.1 to 0.2 yields more accurate throw distances.

5. For fender vaults, the pedestrian-to-vehicle coefficient of friction and the coefficient of restitution are influential in the lateral throw distance, and these parameters can be treated as optimizing parameters for simulations of this trajectory type. Additional work would be needed to establish the likely range on these inputs for this trajectory type.

Case Studies Using the PC-Crash Multibody Pedestrian Model

Case Study: Multibody Simulation to Determine Vehicle Impact Speed

A nighttime crash involving a middle-aged, female pedestrian and two passenger vehicles was reconstructed. The collision occurred on a road in a major city in the Pacific Northwest. One of the involved motorists was driving a Nissan Leaf southbound in a center two-way left-turning lane, approaching an intersection intending to turn left. Another motorist was stopped in a Hyundai Ioniq in the southbound through lane. Traffic was backed up in the southbound lane due to a red signal at the traffic light ahead. A witness in a Ford minivan was stopped in the southbound through lane behind the Hyundai Ioniq. The pedestrian reportedly began crossing the roadway from west to east by walking between the stopped Ford and Hyundai. She continued into the left turn lane where a collision occurred between her and the Nissan Leaf. The pedestrian then contacted the rear driver's side corner of the Hyundai and came to rest on the ground adjacent to this vehicle. Investigation by the police department concluded that the pedestrian was crossing the roadway approximately 200 ft north of the crosswalk. At the time of the collision, the roadway was dry, the conditions were overcast, and it was dark with streetlights present in the area. The speed limit was 25 mph.

Available materials for the investigation included the traffic collision report (TCR), on-scene photographs, dashcam footage from the vehicles of responding officers, and statements and deposition testimony from the involved parties and witnesses. The witness in the Ford minivan stated that the pedestrian walked in front of her vehicle before attempting to cross the center lane. She stated that the Nissan contacted the pedestrian immediately after the pedestrian stepped into the center lane. The driver of the Nissan testified that she stopped her vehicle immediately after striking the pedestrian. She testified that the crash occurred after sunset, it was not raining, and her headlights were on and working. She drove into the center turn lane and passed between one and four cars on her right before the collision occurred. She said the traffic signal at the intersection ahead was initially red. The green arrow came on as she entered the center turn lane. She was in the turn lane for only a short time when she saw a "shadow or a dark form in [her] right peripheral vision." She immediately stopped, fearing that she had struck someone. When she exited her vehicle, she saw the pedestrian lying on the road about one to two car lengths behind her. Following the collision, the Nissan had a large dent over the passenger side front wheel and a dent on the passenger side frame near the windshield. The passenger side mirror was damaged and hanging down. She recalled there was a white or beige smudge on the front passenger side bumper. She did not know if that was related to the subject crash or not. She further testified that the pedestrian was wearing black pants, boots, and a jacket.

Dashcam footage depicted the pedestrian's rest position and clothing. This footage confirmed that the pedestrian was wearing black clothing. The dashcam footage and the on-scene photographs also depicted the rest positions of the involved vehicles. Several on-scene photographs were analyzed to produce the evidence diagram included in **Figure 3.25**.

Figure 3.25 Evidence diagram.

Figure 3.26 depicts the deformation of the passenger's side front fender and the base of the passenger's side A-pillar, along with the damage to the side mirror housing of the Nissan. These areas of damage are from direct contact between the Nissan and the pedestrian. This Nissan Leaf had the S trim package and was equipped with halogen headlights. Based on manufacturer specs, this vehicle weighed approximately 3600 lb, including the driver. The vehicle was equipped with an EDR. However, the vehicle had not been preserved, and the nature of the collision was such that an event was unlikely to be triggered and recorded on the EDR.

Figure 3.26 Damage to the Nissan.

© SAE International

According to the owner's manual, the Nissan Leaf was equipped with an Approaching Vehicle Sound for Pedestrians (VSP) system. The VSP system "is a function that uses sound to alert pedestrians of the presence of the vehicle when it is being driven at a low speed. When the vehicle starts to move, it produces a sound. The sound stops when the vehicle is traveling more than 19 mph (30 km/h) while accelerating. The sound starts when the vehicle speed is less than 16 mph (25 km/h) while decelerating. The sound stops when the vehicle stops." The Nissan was also equipped with AEB, but the AEB system was not equipped with pedestrian detection. Pedestrian detection was an option for the AEB system on other trim packages of the same model year Nissan Leaf, but not with the S trim package. According to the manual, "the AEB system uses a radar sensor located on the front of the vehicle to measure the distance to the vehicle ahead in the same lane…If a risk of a forward collision is detected, the AEB system will provide an initial warning to the driver by both a visual and audible alert. If the driver applies the brakes quickly and forcefully after the warning, and the AEB system detects that there is still a possibility of a forward collision, the system will automatically increase the braking force. If the driver does not take action, the AEB system issues the second visual warning (red) and audible warning and also applies partial braking. If the risk of a collision becomes imminent, the AEB system applies harder braking automatically." The owner's manual noted that the baseline AEB system does not detect pedestrians, and the radar sensor does not detect "pedestrians, animals, or obstacles in the roadway, oncoming vehicles, and crossing vehicles." Thus, the AEB system on the subject Nissan Leaf would not have responded or intervened during the subject incident. This means that the Nissan Leaf came to rest after the subject collision because of braking by the driver, not because of any intervention by the vehicle. In addition, the vehicle would not have given the driver any pre-collision warnings. The Nissan Leaf was also equipped with brake assist. According to the owner's manual, "when the force applied to the brake pedal exceeds a certain level, the Brake Assist is activated generating greater braking force than a conventional brake booster even with light pedal force."

In analyzing the speed of the Nissan at the time of the collision, the distance traveled by the vehicle and the pedestrian from the area of the collision to their points of rest was examined. The pedestrian crossed the southbound lane between the Hyundai Ioniq and the Ford Freestar (there was an approximate eight-foot gap between these vehicles). Based on the reconstructed location of these vehicles, the

pedestrian was crossing the street between 167 and 175 ft north of the crosswalk. The collision occurred when the pedestrian emerged from this area and entered the turn lane. From the area where the collision occurred, the Nissan traveled between 24 and 32 ft to its point of rest. The pedestrian traveled approximately 10 to 11 ft beyond the rear of the Hyundai. Thus, from where she was contacted to rest, she traveled between 10 and 19 ft.

For forward projections and wrap trajectories, the distance traveled by the pedestrian from the impact to rest (the throw distance) can be used to calculate the speed of the vehicle. However, not all types of car–pedestrian interactions are conducive to this analysis. In this case, the pedestrian was initially impacted by the passenger's side front corner of the Nissan. Her body wrapped over the vehicle, her head and/or upper extremities struck the base of the A-pillar, and her mid- to lower body struck the passenger side front fender. As she then rolled off the side of the vehicle, she struck the passenger side mirror. Therefore, this collision would be defined as a fender vault. Since throw distance equations are not applicable to this trajectory type, the PC-Crash pedestrian model was used to simulate the subject collision. According to her driver's license, the pedestrian was 5 ft, 7 in. tall and weighed 150 lb. A multibody pedestrian was scaled to this height and weight. The pedestrian and vehicle models for the simulations are depicted in **Figure 3.27**.

Figure 3.27 Pedestrian and Nissan models in PC-Crash.

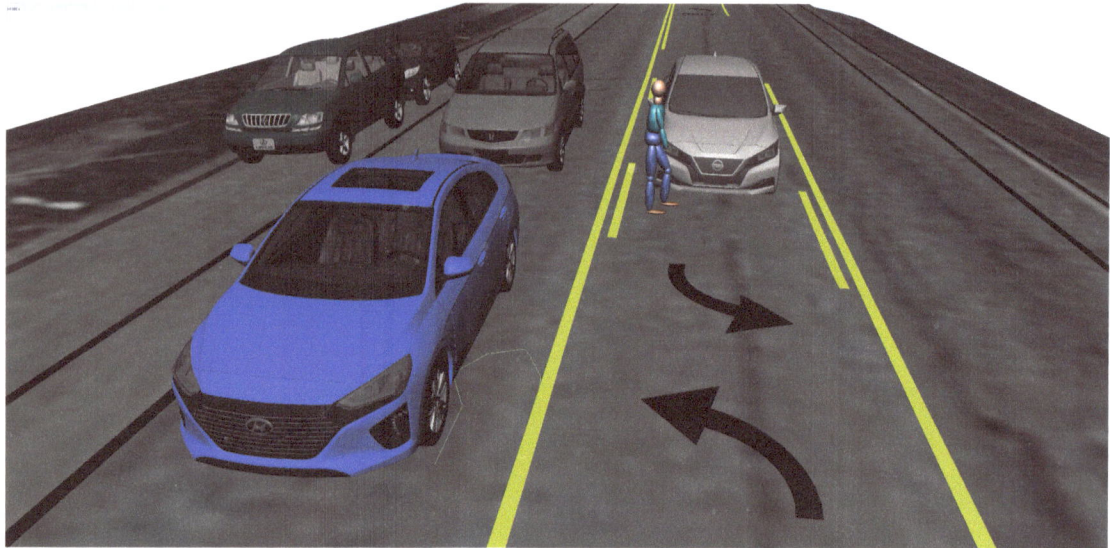

In this case, the PC-Crash pedestrian model enabled the evaluation of the impact conditions that would result in the correct damage to the Nissan and rest position for the pedestrian. In conducting this analysis, a vehicle model was used that accurately modeled the Nissan involved. To validate the model of the Nissan, laser scan data from an exemplar Nissan Leaf was utilized. The impact between the Nissan and the pedestrian was simulated to determine the probable collision speed of the Nissan. In setting up these simulations, a walking speed for the pedestrian of 2.8 mph was used. The speed of the Nissan was varied between 10 and 25 mph in 2.5-mph increments. This analysis resulted in the conclusion that the Nissan was traveling at approximately 15 mph when the collision occurred. This is the speed that produced simulations most consistent with the damage to the Nissan (correct contact points between the pedestrian and the vehicle) and most consistent with the pedestrian's rest position (**Figure 3.28**). Nissan speeds higher than 15 mph resulted in the pedestrian contacting too high on the A-pillar of the Nissan or in the pedestrian striking the Hyundai with too much speed and force. Speeds lower than 15 mph resulted in inadequate contact with the side mirror and not matching the pedestrian rest position. **Figure 3.29** depicts the initial contact with the pedestrian in the simulation that best matched the evidence. **Figure 3.30** shows the pedestrian contacting the fender and A-pillar. **Figure 3.31** depicts contact between the pedestrian and the side mirror.

Figure 3.28 Rest positions for simulation (Nissan = 15 mph).

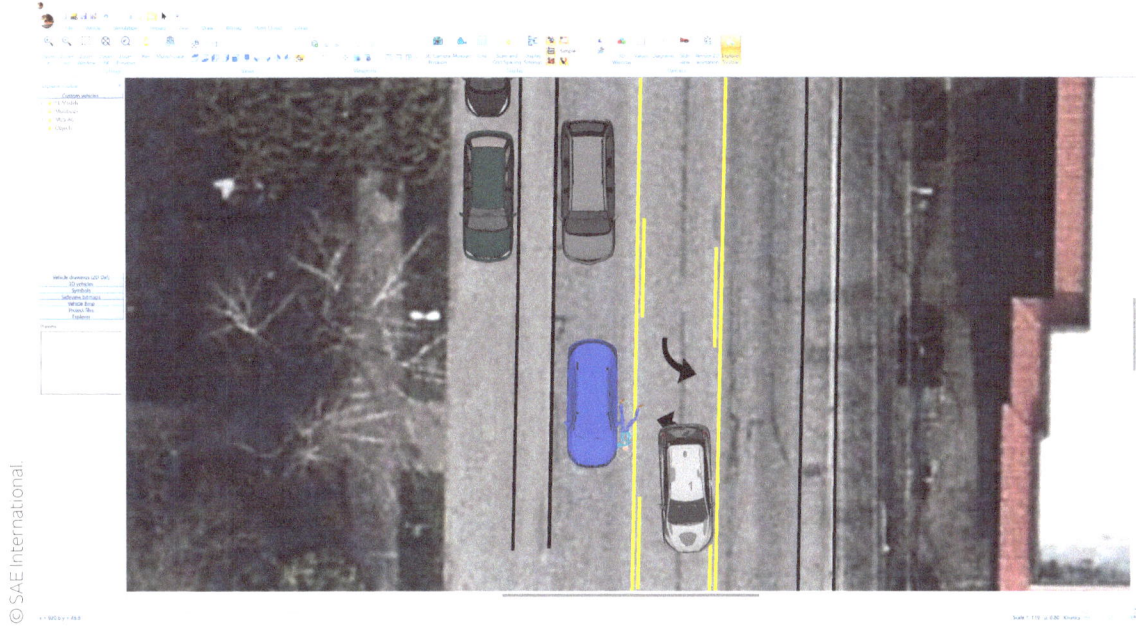

Figure 3.29 Initial contact with the pedestrian.

Figure 3.30 Contact between the pedestrian and the fender and A-pillar.

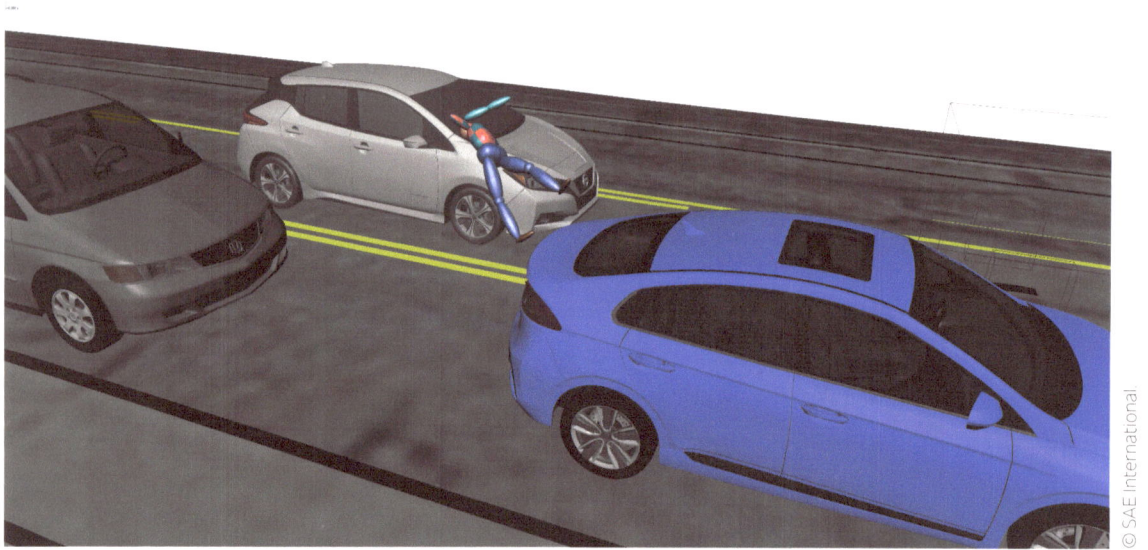

Figure 3.31 Contact with the side mirror.

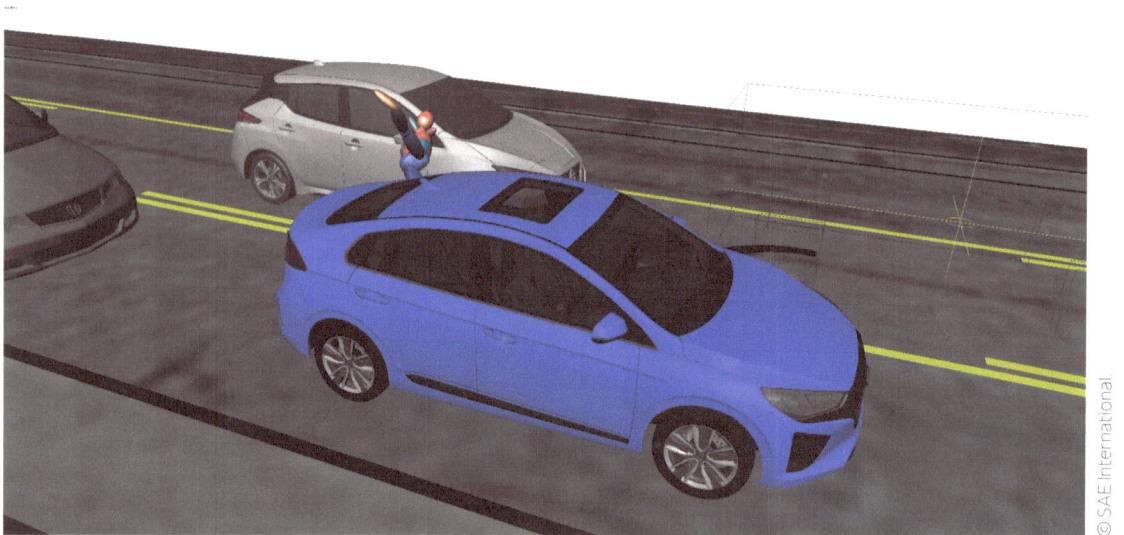

Case Study: Multibody Simulation to Determine Impact Location

Sometimes, the documented physical evidence does not establish where on the roadway a collision occurred. And yet, the collision location can be an important issue for reconstruction. Perhaps the reconstructionist is being asked to determine whether a pedestrian was utilizing an available crosswalk when they were struck. This next case study demonstrates how the impact location can be determined when there is not decisive physical evidence to establish this location. This case study also revisits what can and cannot be determined based on debris from the collision and based on the resting location of objects being carried by the pedestrian. This crash occurred at night and involved a Honda Odyssey minivan striking a middle-aged female pedestrian. The pedestrian was fatally injured. The collision occurred in a rural area in the Pacific Northwest near an intersection without a marked crosswalk. The pedestrian was returning home from work and had ridden the bus. She exited the bus at the bus stop to the north of the intersection and then attempted to cross the roadway. An issue in the case was where the pedestrian had attempted to cross the road. At the time of the collision, it was raining and the roadway was wet. The speed limit was 45 mph. The pedestrian, who would have been crossing from passenger's side to driver's side relative to the vehicle, was struck by the driver's side front corner of the Odyssey and was thrown into the oncoming lane. In the oncoming lane, the pedestrian was struck a second time by the lower driver's side front corner of an oncoming Honda Civic.

According to the statement of the driver of the Odyssey, she remembered seeing a flash of red to her right and then realized there was a person in her lane. She had her windshield wipers on at the time of the collision. She also indicated that there was fog present. There was not a car in front of her. She thinks there was a car behind her. She did not know how far back. According to the statement of the Honda Civic driver, "I just happened to see an object in the air and then it slid on the road. I went to swerve, and I felt a hit, and then recognized that it was a person…we swerved and then we heard a thud, and then went off to the side of the road…All I saw was a person in the air and then sliding. And then I avoided or tried to go to the right." The pedestrian was sliding at an angle toward them. He estimated that he was traveling at 50 mph and indicated that his left tire contacted the pedestrian.

On-scene police photographs revealed collision damage to the driver's side front of the Honda Odyssey, including damage to the hood at the left front corner, the left headlight assembly, the left front fender, the A-pillar, the bumper fascia, and the windshield. The A-pillar was dented above the side mirror, with a piece of scalp and hair lodged between the window trim and pillar. The left side mirror housing had separated from the vehicle, and the glass mirror was missing. The airbags did not deploy. The oncoming Civic exhibited "a light purple smear on the bottom, outside edge of the left bumper cover, in front of the wheel well. There was a drop of blood at the bottom of the left, front wheel hubcap, with a small amount of dried and smeared blood on the hubcap edge and six o'clock position of the tire sidewall. There was no further evidence of contact along the left side or underside of the vehicle."

The pedestrian was 5 ft, 7 in. tall and weighed 225 lb. Investigating officers documented her rest position following the crash. Officers also reported that the pedestrian's bag was at rest near her feet. According to the investigating officers, the pedestrian "was wearing a dark purple jacket and dark blue jeans. A brown leather hiking style boot was present on her right foot. The matching left boot was dislodged and lying to the northwest on its left side in the west portion of the center turn lane. There was a large black and red decorative bag on the north side of the intersection." According to the autopsy report, the pedestrian had a laceration to the back of her head. The autopsy report also stated that the pedestrian had fractures of her left tibia and fibula. The location of these injuries was consistent with the Odyssey striking the pedestrian when her left side was exposed to the vehicle and she was walking from right to left relative to the vehicle. Her left leg was struck first, and then her body rotated such that the back of her head struck the A-pillar, depositing the piece of scalp and hair.

An investigating detective documented the scene using a FARO Focus3D laser scanner; then, he created a scene diagram depicting the scan data and the rest positions of the Hondas and the pedestrian. The diagram also depicted the detective's estimate of where the collision occurred. The detective's diagram showed the collision occurring with the pedestrian crossing at the intersection. The detective reported: "There are several pieces from the minivan's headlight that show up in the southbound lane on the north edge of the intersection...There is a handbag and a bagel bag near the center line of the northbound lane. This was their approximate final rest location...I looked but did not see any pre- or post-impact tire marks associated with the minivan. I did not find the scuff marks associated with the pedestrian." Another detective stated: "The debris field was primarily contained in the SB and two-way center turn lanes, within the intersection. It appeared to 'V' out to the east and south from the left portion of the SB lane."

It is important to recognize that vehicle debris from a pedestrian collision is not a reliable indicator of the location of impact. Debris is often thrown, projected, or carried downstream during the collision. As Fricke [3.14] has noted, "debris from a car can be a considerable distance from the first contact positions." The PC-Crash multibody model provided a way forward for the analysis of this crash. In this case, this model enabled the evaluation of impact conditions that resulted in the correct rest positions, damage exhibited by the vehicles, and injury locations to the pedestrian. In conducting this analysis, vehicle models were used that accurately modeled the vehicles involved. For example, to validate the model of the Odyssey, laser scan data from an exemplar Odyssey was used. To prepare a scene model for use in the simulations, publicly available lidar data from the United States Geological Survey's (USGS) 3D Elevation Program (3DEP) was obtained. The diagram created by the investigating detectives was overlaid with the USGS lidar so that the roadway evidence documented by the detectives was incorporated into the scene model. The PC-Crash scene that was developed and used for the analysis is viewable in **Figure 3.32**.

Figure 3.32 Pedestrian motion resulting from Odyssey impact.

The impact between the Odyssey and the pedestrian was simulated to determine the probable area of impact, the pre-collision travel direction of the pedestrian, the speed of the Odyssey, and the speed of the pedestrian. The collision between the pedestrian and the Civic was also simulated. In setting up these simulations, a range of speeds for the pedestrian between 2.5 and 6.9 mph was utilized. There were no witnesses who observed the pedestrian prior to her being struck, so it was unknown whether she was walking or running. Variations in the pedestrian's travel direction relative to the Odyssey and the impact speed of the Odyssey were also examined. In developing these simulations, the Honda Civic driver's statements that he was traveling 50 mph and that he saw the pedestrian land on the roadway and slide prior to impacting his vehicle were utilized. Simulations were run varying parameters such as the initial impact location on the roadway and the orientation of the pedestrian model at the impact. The degree to which each simulation matched the physical evidence was then evaluated. The best match with the physical evidence was obtained with the pedestrian traveling a walking speed diagonally across the roadway and angled slightly away from the Odyssey at impact. The simulations also showed that the pedestrian was crossing to the north of the intersection. The impact location was approximately 35 to 45 ft farther north than what the investigating detectives assumed. The Odyssey was traveling approximately 40 mph at impact. **Figure 3.32** presents a screen capture from PC-Crash showing the motion of the pedestrian that resulted from impact with the Honda Odyssey in the simulation that best matched the evidence.

The process for analyzing this collision can be summarized as follows:

1. The physical evidence was documented. This included:

   - Damage locations on the Odyssey.
   - Injury locations on the pedestrian.
   - Damage locations on the Civic.
   - Rest positions for the Odyssey, the Civic, and the pedestrian.

2. Physical constraints were established based on the statements by the driver of the Civic:

   - The pedestrian was airborne when he first observed her.
   - The pedestrian landed and slid on the ground into impact with the Civic.
   - The pedestrian was approaching the Civic at an angle.

3. An accurate three-dimensional scene was created in PC-Crash.

4. Accurate DXF models were imported into PC-Crash for both the Odyssey and the Civic.

5. The PC-Crash pedestrian model was scaled to the correct height and weight for the pedestrian.

6. Ranges were established for the inputs into the simulations:

   - The speed of the Odyssey.
   - The speed of the pedestrian.
   - The travel direction of the pedestrian.
   - The crossing location of the pedestrian.
   - The body posture of the pedestrian.
   - The friction and restitution inputs for the pedestrian models.
   - The speed of the Civic.

7. Simulations were run with many combinations of these variables, and the simulations were evaluated based on their fit with the physical evidence and the physical constraints established from the Civic driver's statement.

8. Conclusions were reached based on the simulations that exhibited a good fit with the physical evidence and physical constraints.

Case Study: Multibody Model in PC-Crash to Determine Bicyclist's CG Location

This case study demonstrates using the PC-Crash pedestrian model other than simulating a collision. A crash was reconstructed that involved a bicyclist colliding with a pedestrian. The bicyclist was traveling southbound in a bicycle lane, and the pedestrian attempted to walk westbound through the bicycle lane in a crosswalk. The crash occurred during daylight hours. The weather was overcast, and the sidewalk and bike path were dry. As the bicyclist was approaching the crosswalk, the pedestrian stepped into the crosswalk. The bicyclist responded with aggressive braking utilizing both brakes, and this led to him pitching over the handlebars of his bicycle. The bicyclist and his bicycle struck the pedestrian while he was in the midst of pitching over.

The location of this collision was inspected, mapped, and photographed. The mapping utilized a FARO laser scanner and a DJI Mavic 2 Pro sUAS. GPS-enabled Propeller AeroPoints were placed at the site as ground control. After the inspection, Pix4D Mapper photogrammetry software was used to generate a three-dimensional map and point cloud of the crash site using the aerial photographs. The data were supplemented with the three-dimensional point cloud generated with the FARO laser scanner.

Using the mapping data, the downgrade of the bicycle lane was measured at approximately 4.5% (2.6°) in the area of the collision. According to the crash report, the collision occurred in a crosswalk just to the south of a bus stop shelter, and then, the bicyclist and his bicycle came to rest in another crosswalk at the intersection just to the south of the bus stop. Based on the mapping data, the distance from the center of the crosswalk where the collision occurred to the center of the crosswalk where the bicyclist came to rest was approximately 21.5 ft. The distance from the south edge of the crosswalk where the collision occurred to the north edge of the crosswalk where the bicyclist came to rest was approximately 12.5 ft. The width of the crosswalk and bicycle lane where the collision occurred was approximately 5 ft.

Provided bodycam footage from investigating officers included the following frame

(**Figure 3.33**). This frame looks southbound from a position just to the north of the bus stop shelter, along the direction the bicyclist was traveling prior to the collision. This image depicts a number of skid marks from bicycles preceding the subject crosswalk. One of these skid marks, which is identified with red arrows in the figure, is darker than the rest and was consistent with the heavy braking described by the bicyclist and the pitch-over motion that resulted. There were three possibilities related to this skid mark: (1) it was from the subject crash, and it was deposited by the front tire of the bicycle; (2) it was from the subject crash, and it was deposited by the rear tire of the bicycle; and (3) it was not from the subject crash. In analyzing this collision, each of these possibilities was considered. Based on the analysis of this frame and others, the skid mark was approximately 20 ft long. It began around 25 ft north of the crosswalk.

Figure 3.33 Frame from bodycam footage.

The bicyclist's helmet, backpack, and bicycle were also captured in the bodycam footage. The bicycle was a Turner Sultan equipped with 29er Bontrager Wheels and Schwalbe Nobby Nic Tires. Using manufacturer specifications and online sources, the weight of the bicycle was calculated at approximately 27 lb. The bicyclist reportedly weighed 180 lb, and he was 6 ft, 1 in. tall. The combined weight of his helmet, backpack, and cargo was estimated at approximately 10 lb. When evaluating the motion of two-wheeled vehicles, bicycles are different from motorcycles by having a rider who weighs significantly more than the vehicle itself. This means that the rider's body will have a significant influence on the handling of the bicycle and will significantly influence the level of braking that will cause the bicycle to pitch over. Quantifying this influence involves determining the longitudinal and vertical CG positions for the bicycle–rider combination. To determine the likely CG position for the bicyclist and his bicycle, the multibody model in PC-Crash was used. One of the stock multibody mountain bicycle and rider models from PC-Crash is depicted on the left side of **Figure 3.34** (*diamond rigid with rider 20171127.mbdef*). This mountain bike model consists of 40 rigid bodies attached by joints. Each of the wheels (including the axles) is modeled with 13 bodies. The frame is modeled with ten bodies, the fork with two bodies, the seat with a single body, and the handlebars with a single body. This stock bicycle model has a weight of 33.1 lb, a wheelbase of 44.2 in., a seat height of 36 in., and wheel diameters of 27.6 in. The size, weight, and position of each of the bodies can be altered by the user, and this capability was used to create a custom bicycle model that more closely matched the characteristics of the subject bicycle (see the right side of **Figure 3.34**). The rider was also scaled to match the height and weight of the subject bicyclist.

Figure 3.34 Stock bicycle/rider model (left) and custom model (right).

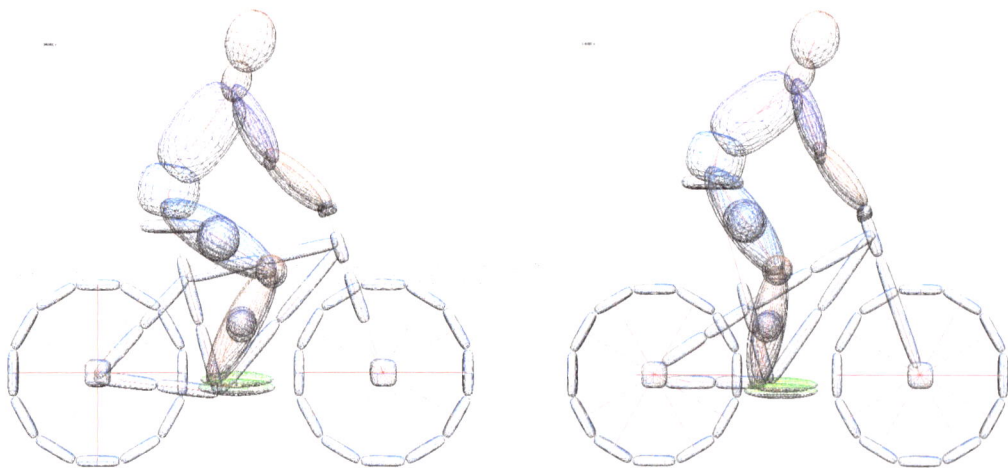

This modeling led to the conclusion that the combined CG location of the bicyclist and his bicycle was approximately 25 to 26 in. behind the front axle and approximately 43 to 44 in. above the ground. With this position for the combined CG, a pitch-over would be produced if the bicyclist braked hard enough to generate 0.6g of deceleration. This calculated value neglects the compression of the front suspension during heavy braking, which would lower the pitch-over threshold. The actual pitch-over threshold was likely to be in the range of 0.55 to 0.58g. This is consistent with decelerations reported in the literature to produce a pitch-over of a bicycle and rider [3.15].

In conducting the analysis, walking speeds for the pedestrian between 2.0 and 4.5 mph (3.0 to 6.6 ft/s) were considered. At these speeds, it would have taken the pedestrian between 0.8 and 1.7 sec to travel through the five-foot bicycle lane. These times assume a constant speed. If the pedestrian had started from a stop, it would take her approximately one foot to accelerate up to a speed of 2 mph. This would take 0.65 sec. It would then take her an additional 1.36 sec to travel the additional 4 ft of the bicycle lane. Thus, with the start-up time, it would take the pedestrian approximately 2 sec to cross at 2 mph. It would take the pedestrian 4.8 ft to accelerate up to 4.5 mph. This would take approximately 1.47 sec. It would then take her 0.03 additional seconds to finish traversing the bicycle lane. Thus, with the start-up time, it would take the pedestrian approximately 1.5 to 2 sec to traverse the five-foot bicycle lane.

The PC-Crash single-track vehicle driver model [3.16] was used to analyze the motion of the bicyclist and his bicycle in the moments leading up to and following the subject collision. This model enabled the examination of the braking levels and timing necessary for the bicycle to deposit the 20-foot-long skid and then to pitch over and slide to the rest position. In conducting this analysis, simulations were generated in which either the front or rear tire of the bicycle deposited the skid mark and the bicycle pitched over and came to rest in the correct location. The time it takes for a vehicle operator to implement an emergency response to an immediate hazard is referred to as the perception-response time. Data from the literature revealed that vehicle operators confronted with a path intrusion similar to what the bicyclist was confronted with would enact their emergency response within approximately 0.9 sec. The skid mark began approximately 28 ft prior to the center of the crosswalk. If he was initially traveling at 15 mph and his rear tire deposited the skid mark, it would take the bicycle around 1.45 sec to reach the middle of the crosswalk. Adding a perception-response time to this, the bicyclist would have to begin responding to the pedestrian approximately 2.15 sec prior to the collision. This is prior to the time that the pedestrian entered the bicycle lane, indicating that if the skid was from the subject crash, the bicyclist was anticipating the pedestrian stepping into the crosswalk. This may be an indication that she did not stop prior to entering the crosswalk. Similarly, if the bicyclist was initially traveling at 15 mph and his front tire deposited the skid mark, it would take the bicycle around 2.1 sec to reach the middle of the crosswalk. Adding a perception-response time to this, the bicyclist would have to begin responding to the pedestrian approximately 2.8 sec prior to the collision, again prior to the time the pedestrian entered the bicycle lane. Several video frames of this collision were later discovered which revealed that the tire mark was from the subject crash, that it was deposited by the rear wheel, and that the pedestrian did not stop prior to stepping into the crosswalk.

References

3.1. Rose, N., Smith, C., Carter, N., and Metanias, A., "Validation of Pedestrian Collision Reconstruction Using the PC-Crash Multibody Pedestrian Model," SAE Technical Paper 2025-01-8681, doi:https://doi.org/10.4271/2025-01-8681.

3.2. Moser, A., Steffan, H., and Kasanický, G., "The Pedestrian Model in PC-Crash – The Introduction of a Multibody System and Its Validation," SAE Technical Paper 1999-01-0445 (1999), doi:https://doi.org/10.4271/1999-01-0445.

3.3. Moser, A., Hoschopf, H., Steffan, H., and Kasanicky, G., "Validation of the PC-Crash Pedestrian Model," SAE Technical Paper 2000-01-0847 (2000), doi:https://doi.org/10.4271/2000-01-0847.

3.4. Moser, A., Steffan, H., and Strzeletz, R., "Movement of the Human Body versus Dummy after the Collision," in *Proceedings of the 1st Joint ITAI-EVU Conference, 18th EVU Conference, 9th ITAI Conference*, Hinckley, UK, 2009, 87-105.

3.5. Wach, W. and Unarski, J., "Fall from Height in a Stairwell – Mechanics and Simulation Analysis," *Forensic Science International* 244 (2014): 136-151.

3.6. Richardson, S., Josevski, N., Sandvik, A., Pok, T. et al., "Pedestrian Throw Distance Impact Speed Contour Plots Using PC-Crash," SAE Technical Paper 2015-01-1418 (2015), doi:https://doi.org/10.4271/2015-01-1418.

3.7. Piloto, P. and Teixeira, V., "Pedestrian Forward Projection after Vehicle Collision," *Journal of Mechanical Engineering and Biomechanics* 2, no. 5 (2018): 75-81.

3.8. Tilley, A.R. and Henry Dreyfuss Associates, *The Measure of Man and Woman: Human Factors in Design* (New York: John Wiley & Sons, Inc., 2002).

3.9. Fugger, T., Randles, B., Wobrock, J., and Eubanks, J., "Pedestrian Throw Kinematics in Forward Projection Collisions," SAE Technical Paper 2002-01-0019 (2002), doi:https://doi.org/10.4271/2002-01-0019.

3.10. Toor, A. and Araszewski, M., "Theoretical vs. Empirical Solutions for Vehicle/Pedestrian Collisions," SAE Technical Paper 2003-01-0883 (2003), doi:https://doi.org/10.4271/2003-01-0883.

3.11. Severy, D. and Brink, H., "Auto-Pedestrian Collision Experiments," SAE Technical Paper 660080 (1966), doi:https://doi.org/10.4271/660080.

3.12. Carter, N., Hashemian, A., and Mckelvey, N., "An Optimization of Small Unmanned Aerial System (sUAS) Image Based Scanning Techniques for Mapping Accident Sites," *SAE Int. J. Adv. & Curr. Prac. in Mobility* 1, no. 3 (2019): 967-995, doi:https://doi.org/10.4271/2019-01-0427.

3.13. Becker, T., Reade, M., and Scurlock, B., "Simulations of Pedestrian Impact Collisions with Virtual Crash 3 and Comparisons with IPTM Staged Tests," *Accident Reconstruction Journal* (2016).

3.14. Fricke, L.B., *Traffic Crash Reconstruction*, 2nd ed. (Evanston, IL: Northwestern University Center for Public Safety, 2010).

3.15. Broker, J. and Hottman, M.M., *Bicycle Accidents, Crashes, and Collisions: Biomechanical, Engineering, and Legal Aspects*, 2nd ed. (Tucson, AZ: Lawyers and Judges Publishing Company, 2017), ISBN:978-1-936360-58-1.

3.16. Palmer, J., Rose, N., Smith, C., Walter, K. et al., "Validation of the PC-Crash Single-Track Vehicle Driver Model for Simulating Motorcycle Motion," SAE Technical Paper 2024-01-2475 (2024), doi:https://doi.org/10.4271/2024-01-2475.

Chapter 04

Video Analysis

Nathan Rose and Connor Smith

Pedestrian crashes are sometimes captured by a surveillance camera attached to a nearby business, by a traffic monitoring camera, or by a camera attached to the dash or windshield of a vehicle. When a fixed surveillance or traffic camera captures a crash, the camera position, orientation, and field of view are often fixed, and the reconstructionist will apply video analysis methods to track the motion of the vehicles and pedestrians depicted in the footage. When a dash- or windshield-mounted camera captures the crash, the camera will be moving along with the vehicle. In these instances, the analysis will begin by tracking the motion of the camera and then proceed to tracking the motion of the vehicles and pedestrians within the view of the camera.

Video analysis and tracking often begins when footage is provided or obtained. Before beginning to analyze the footage, it is a good idea to make a copy of the provided or obtained footage and to perform the analysis on the copy. This ensures that an original version of the video is maintained and does not become altered by the analysis process. After making a copy, the analysis will proceed to evaluating the characteristics of the video—the aspect ratio, resolution, frame rate, and structure. During this process, individual frames of the footage are often exported for analysis. These individual frames are analyzed to verify metadata information and characteristics, establish the camera field of view, and remove lens distortion. Then, a suitable analysis method is selected. In recent years, video analysis has become an integral part of accident reconstruction. This chapter covers some of the important principles and methods but does not attempt to be exhaustive. As with any of the chapters in this book, the reader may benefit from also studying the cited literature.

Gate Analysis

Gate analysis is a simple method for tracking the motion of a camera or the motion of vehicles and pedestrians depicted in video footage. In this method, a series of gates is established using landmarks visible in the video. Sometimes these landmarks are physical objects on the road surface—a stop bar, a lane marking, or some other striping on the road, for example. This is common when the camera is attached to a vehicle and is moving. In other instances, when the camera is attached to a building, lines are projected out from the camera location, through landmarks visible in the scene (trees and poles, for example), and out onto the road (or other surface the objects are moving along). This establishes the gates on the road surface, and the distance between these gates can be measured. The analyst then establishes the video frames at which the vehicle being tracked reaches each gate. The number of frames for the vehicle to travel between two gates then establishes the time it takes for the vehicle to travel between gates. The average speed of the vehicle or

pedestrian between gates can then be calculated by dividing the distance traveled by the time to travel that distance.

This method can yield an adequate level of accuracy. However, it is important to keep in mind that the calculated speed is an average between gates. Often, the number of available landmarks in the video is limited, and the distance between gates can be large. On the one hand, this minimizes the uncertainty in speed calculation because uncertainties in distance and frame times have a smaller influence over a longer distance. On the other hand, this method will miss any fluctuations in the speed of the vehicle between gates. If the vehicle is accelerating or decelerating, this can be missed with large distances between gates. There may also be uncertainty related to the lateral position of a tracked vehicle (within a lane, for example), and a range of positions at each gate might need to be considered.

The gate method starts with extracting the individual frames of the video, which yields a series of still frames for analysis. A subset of the available frames is then selected for analysis. Typically, during the process of extracting individual video frames, the reconstructionist will determine the frame rate of the video. The frame rate is the number of individual frames that are captured during each second of the footage. Some cameras record with a constant frame rate and others have a variable frame rate. For constant frame rate cameras, the frame rate can be used to calculate the duration of the interval between any two selected frames. For a variable frame rate camera, additional analysis will be needed. If the time between selected frames is known or can be established, then the average speed of the camera and the tracked objects can be determined by measuring the distance traveled by the camera or tracked objects between the frames and dividing the distances by the duration of the interval between these frames. An example of the application of this method is presented next.

Case Study: Gate Analysis for a Vehicle Approaching a Pedestrian

A pedestrian collision was reconstructed that occurred at around 6 a.m. in May in a city in the Pacific Northwest. According to the police investigation, the collision occurred when a female pedestrian was walking across the road in a midblock crosswalk between a parking garage on the west side of the street and her place of employment on the east side of the street. While in the crosswalk, the pedestrian was struck by a Chrysler sedan that was travelling southbound. On this day, sunrise occurred at 5:35 a.m., and so the collision occurred under daylight conditions. The roadway was dry, and the weather was clear. The speed limit was 30 mph.

Figure 4.1 is an aerial photograph depicting the area of the crash, oriented such that north is up on the page. As this image shows, there was a southbound lane, a northbound lane, and a center turn lane. The painted crosswalk that traverses the roadway is depicted in the lower portion of the aerial. Parking lanes are present at various locations along the roadway. There were several motion-activated surveillance cameras in the area that captured footage relevant to this

reconstruction. The actual collision occurred out of the view of any of the cameras, but the Chrysler was captured as it approached the crosswalk. The footage was provided as an executable file. Double-clicking this file would open seven streams of video footage that played in a proprietary viewer. Two of the video streams showed the Chrysler as it approached the crosswalk. **Figure 4.1** identifies the locations of these cameras with yellow circles. The first camera was around 550 ft north of the crosswalk, and the second camera was around 170 ft north of the crosswalk. **Figure 4.2** is a sample frame from the northmost camera that depicts the Chrysler. This camera looks south. **Figure 4.3** is a sample frame from the south camera that also depicts the Chrysler. This camera looks north.

Frames from these cameras were reviewed, and frames were selected in which the Chrysler could be located relative to striping on the roadway and a sign. These were landmarks that were also visible in the aerial imagery. Google Earth Pro was used to measure the distance (228 ft) between these landmarks. The timestamp on the selected frame from the north camera was 5:56:40.887, and for the selected frame from the south camera, it was 5:56:47.207. Thus, based on the timestamp printed on the frames, 6.32 sec elapsed as the Chrysler traveled between the depicted positions. Based on the distance traversed by the Chrysler and the elapsed time, the Chrysler was traveling at an average speed of approximately 25 mph.

Figure 4.1 Aerial photograph showing camera locations.

Figure 4.2 Frame from northmost camera looking south.

Figure 4.3 Frame from south camera looking north.

Camera Matching Video Analysis

Camera matching photogrammetry is another commonly used method for tracking the motion of a camera and objects within the view of the camera. This method involves first reconstructing the position, orientation, and field of view of the camera that captured a frame and then using these known camera characteristics—combined with data about the geometry of the road surface—to locate a vehicle or other object within the view of the camera. Similar to gate analysis, camera matching video analysis begins with extracting individual frames from the video and determining the frame rate of the video. Different from the gate method, camera matching is not dependent on the tracked vehicle or pedestrian passing identifiable landmarks. Instead, this method depends on having identifiable objects or landmarks within the view of the camera that enable the camera position, orientation, and field of view to be reconstructed. If the camera characteristics can be reconstructed, and the geometry of the road surface is known, then vehicles and other objects can be located within the frame, even if they are not located adjacent to identifiable landmarks.

When analyzing video frames from a dash- or windshield-mounted camera, the camera matching process itself usually consists of the following steps:

1. A series of video frames is selected for analysis. If the camera characteristics are known, lens distortion may be removed from the frames immediately after selection of the frames. All camera lenses produce some level of distortion in the resulting image, and this distortion can introduce errors into the camera matching process if it is not removed. Lens distortion occurs in two basic forms—barreling and pin cushioning. Barreling is distortion that squashes the edges of the image and stretches the center of the image. Pin cushioning is distortion that stretches the edges of the image and squashes the center of the image. Generally, barreling occurs from wide focal lengths, and pin cushioning occurs from zoom focal lengths. Lenses can also produce a mix of barreling and pin cushioning. Methods for removing lens distortion are discussed in previous studies [4.1, 4.2]

2. The reconstructionist makes note of static objects depicted by the video frames, such as the road surface, the roadway striping, curbs, signs, trees, and bridges. If these objects still exist at the site, they can then be documented and mapped. This mapping could be carried out with a laser scanner or using image-based scanning methods. If the object no longer exists at the site, documentation and mapping of their original locations can potentially be obtained from historical aerial imagery or publicly available lidar data.

3. The reconstructionist then uses computer modeling software to create a virtual camera corresponding to each frame of video being analyzed. This virtual camera is used to view the mapping data from a perspective that is visually similar to that shown in the video frame.

4. The video frames are then imported into the modeling software, and each frame is designated as a background image for the corresponding virtual camera.

5. The analyst then adjusts the location, field of view, and viewing plane of the virtual camera until an overlay is achieved between the mapping data and the geometry shown in each video frame. If the distortion has not already been removed from the video frames, the distortion correction can be performed in conjunction with the process of achieving the overlay. When this overlay is obtained, the camera location, field of view, and viewing plane can be reconstructed. If the position of the camera relative to the vehicle is known, then the reconstructed camera positions also provide reconstructed positions for the vehicle carrying the camera.

6. Once the camera position, orientation, and field of view are known, objects within the view of the camera can be located within the scene. The process of locating these objects benefits from establishing physical constraints on their location (i.e., the vehicle must be sitting on the roadway surface).

As with any method of video analysis, there will be some level of uncertainty in the speeds of the camera and other objects calculated from the tracked positions. Uncertainty can arise due to uncertainty in locating the position of the camera and uncertainty in the duration of the time interval between frames. If the frame rate of the camera is constant, uncertainty in the frame interval timing can be ignored. Beauchamp, Pentecost et al. [4.3] showed that, for a camera with a known, constant frame rate, the uncertainty in the calculated average speed between two selected frames is given by the following equation:

$$\delta s = \sqrt{\left(\frac{1}{\Delta t}\delta d_1\right)^2 + \left(-\frac{1}{\Delta t}\delta d_2\right)^2} \tag{4.1}$$

where

Δt is the time interval between the frames selected for analysis

δd_1 is the magnitude of the positional uncertainty in the camera or vehicle position at the beginning of the frame interval

δd_2 is the magnitude of the positional uncertainty at the end of the frame interval

This equation assumes that there is no uncertainty in the time interval between frames and shows that the uncertainty in the speed is minimized by lengthening this time interval (more time between analyzed frames). There is a limit, though, to how much the calculation interval can be lengthened without missing peaks or valleys in the speed or changes in the speed that are physically significant. Another approach to minimizing the uncertainty in the calculated speed is to articulate physical constraints that the resulting speeds must meet. For example, if the video analysis is showing that the vehicle that is undergoing emergency-level braking, a constraint on the analysis could be that the implied deceleration of the vehicle from the video analysis cannot exceed 1g (or some similar value based on the braking capabilities of a particular vehicle).

The accuracy and precision of the camera matching method can be influenced by the resolution of the individual video images and the angle of incidence between the camera and the object being located. Terpstra, Hashemian et al. [4.4] explored the influence of resolution and angle of incidence on the accuracy of the camera matching method. They also compared the accuracy achieved with an automated implementation of the camera matching method and a manual implementation. The resolution of an object within an image can be measured based on the number of pixels used to depict the object. The larger the portion of the image the object fills, the higher the number of pixels that will be used to depict the object and the higher the object resolution within the image. Terpstra and Hashemian noted that "higher object resolution will allow for greater accuracy when determining the objects position and orientation." There are limits to this, though. If an object fills the entire video frame, it will be depicted with the maximum possible object resolution, but it will be impossible to determine its position and orientation because there will be nothing else depicted in the image to which its orientation can be referenced.

When the camera matching method is implemented to analyze a video, the video is often the only available view, and so the camera matching method is often applied as a single-image photogrammetric method. As Terpstra, Hashemian et al. [4.4] have noted, "to solve for the location of the evidence using camera matching photogrammetry with only a single image, the surface that the evidence is resting on must be known, documented, and represented within the computer model." When it comes to tracking the motion of a vehicle across a series of video frames, this means that the road surface on which the vehicle is traveling provides a constraint on the analysis of the position of the vehicle. The vehicle must be on the road surface (assuming it is actually on the road surface in the video), and this constrains the analyst's reconstruction of the vehicle location using the camera matching method. "The surface the evidence is resting upon prevents the object from being placed closer or farther from the camera and limits the range of possible solutions."

As noted above, the accuracy of the camera matching method can be improved by removing lens distortion from the video frames prior to aligning them with the scene mapping data. Different camera and lens combinations produce different levels of distortion in the recorded images, and so varying levels of improvement will be obtained from this removal. If the distortion is minimal, the improvement in accuracy will be minimal; if the distortion is significant, the improvement in accuracy can be significant. In a legal setting, where there is concern about alteration or manipulation of raw video frames, the question can arise: How does distortion removal alter the video image and is this an acceptable manipulation of the raw video data?

This concern is first addressed by stating that some tasks within video analysis will utilize the

raw, unaltered video frames. Thus, raw video frames should be maintained and utilized, where appropriate and needed. However, within the context of the camera matching method, removing lens distortion moves the pixels within the image in such a way that the image more faithfully depicts the shape of the real-world scene. The distorted image depicts the real-world scenario in a distorted way that can compromise a reconstructionist's ability to extract accurate measurements from the image. Thus, creating an image that faithfully depicts the real-world situation without distortion is an essential step in the process of aligning the image to the scene mapping data. The technical literature is unanimous that, within the context of camera matching and other forms of photogrammetric analysis, correction for lens distortion is a good and necessary part of the process (see, for example, [4.1] and [4.2]).

Neale et al. [4.1] demonstrated that camera and lens combinations of the same make and model will produce the same distortion in the resulting images. This means that, if a reconstructionist can identify the make and model of the camera and lens an image was captured with, then an exemplar camera and lens can be used to determine the lens distortion characteristics for correcting the images. The distortion characteristics of a camera and lens combination will vary with the focal length. Neale et al. analyzed the lens distortion characteristics for 35 cameras and demonstrated that, for an image with distortion, the center of the image contains the least amount of distortion. The amount of distortion increases moving out from the center, with the maximum distortion occurring at the edges of the image. They also demonstrated that barrel distortion is associated with wide-angle (smaller) focal lengths and pin cushion distortion is associated with zoom (higher) focal lengths. Focal lengths around 50 mm tended to exhibit the least amount of distortion, and it was around this focal length that many of the lenses transitioned from barrel to pin cushion distortion.

A study published in 2021 by Simacek et al. [4.5] examined uncertainties and errors that can arise in the speeds obtained from the tracked positions of vehicles in a video. These uncertainties and errors can be influenced by video resolution, frame rate, lens distortion, motion blur, and camera movement. This study tracked vehicles in video footage obtained from different video systems—three moving cameras and two stationary cameras—and compared the results to the known motion documented with other instrumentation. To represent stationary surveillance cameras, two cameras (a GoPro HERO5 and a Sony α6400) were positioned at the top corner of a business complex. The moving cameras were contained within one of the vehicles involved in a mock collision scenario—a 2018 Tesla Model 3 Dashcam, a dashboard video system with a low frame rate, and a Blackmagic Design camera. These videos were captured as this vehicle traveled toward the other vehicle involved in the mock collision scenario.

The two vehicles involved in the mock accident scenario were driven toward and past each other in adjacent lanes under dry, daylight conditions on a smooth, recently resurfaced asphalt roadway. The vehicle carrying the cameras was traveling at approximately 35 mph, and the approaching vehicle was traveling at approximately 45 mph. The authors noted that "the point where they passed one another was considered the area of impact in the mock accident scenario." In relation to the speed of the approaching vehicle, the authors noted that

"the moving cameras yielded an average percent error of 3.2%. The stationary cameras yielded an average percent error of 1.15%." The percent error varied from one camera to another, and the reader is encouraged to refer to the study for data related to the specific cameras.

From a big picture perspective, these authors concluded that the following factors influenced the resulting error: (1) a moving camera generally produced higher errors than a fixed camera (although there was one exception); (2) images with more initial lens distortion produced higher errors than images with less initial distortion (because some distortion was still present in the images after the distortion removal process); (3) images with the target vehicle moving toward the camera generally produced higher error than images with the target vehicle moving across the view of the camera; and (4) motion blur increased the error. For the moving cameras, the error generally dropped as the approaching vehicle got closer to the camera because the vehicle filled more pixels of the image as it got closer. However, as the approaching vehicle began to pass the camera and the vehicle moved into the portion of the image with the most residual distortion (the edges of the image), the error began to increase again.

Case Study: Camera Matching Analysis to Track Bus Motion

A collision involving an articulated bus and a pedestrian was reconstructed. The collision occurred at an intersection when it was dark and raining, and the asphalt roadway was wet. The bus was being driven northbound in the right toward the intersection. There were two northbound through lanes, which were flanked by left and right turn pockets. Buses could utilize the right turn lane as a through lane, and this was the lane where the bus was operating. The northbound and southbound lanes were separated by a raised median. The driver of the bus had a green light as he approached. A 79-year-old female pedestrian was crossing the intersection in a crosswalk, traveling from west to east on the south side of the intersection (moving from left to right relative to the bus). The pedestrian was reportedly 5 ft, 1 in. tall and weighed 120 lb. She was struck by the left front of the bus. The bus was equipped with cameras, and the collision was captured by the front-facing camera that was situated at the front of the bus.

Approximately 41 million measurement points were collected at the site using two Faro-brand laser scanners to document the roadway and intersection geometries. Mapping data captured by the investigating officers was incorporated into these Faro measurement points. During an inspection of the bus, the cameras were documented. The forward-facing camera was an Apollo RR-CIR236. An employee of Apollo Video Technology indicated that the camera was an analog camera that generated 30 fps. However, the digital video recorder (DVR) on the bus was set up to capture and record 10 fps from each camera stream. However, "the DVR can only handle so much data at a time. There is a limit defined by: # of Cameras × frame rate × resolution. If that limit is exceeded by the DVR's configuration, the frame rate may drop below the configured frame rate. However, it will do so consistently, and it will not vary within the video capture itself. For instance, if the limit is exceeded, and the frame rate is set for 10 frames per second, it may drop to 8.5 frames per second, but it will be consistently 8.5 frames per second,

throughout the video, for all cameras." To determine the frame rate of the footage from the forward-facing camera, we examined the duration of two yellow traffic signals visible in the video after the collision had occurred. The traffic light sequencing sheets for the intersection indicated that the yellow signal duration was 4 sec. This duration yielded an average frame rate for the video of between 8.5 and 8.75 fps. This was the frame rate that would give the yellow signal the correct 4-sec duration.

During the inspection of the bus, we captured footage from the forward-facing camera. We placed a high-speed clock within the view of the camera, along with several objects of known size. We also scanned the portion of the building that was visible within the view of the camera. This documentation enabled us to remove the distortion from the video frames: first, the video frames captured during our inspection and, ultimately, the video frames depicting the subject collision. Removing the distortion from the video images resulted in a series of video frames that accurately depicted the geometry shown by the camera. Lines that were straight in the real world also appeared straight in the distortion-corrected video frames. The next series of graphics illustrated the process of removing the distortion from the video frames. **Figure 4.4** shows a single frame of video from the inspection. **Figure 4.5** shows the aspect ratio correction that was necessary for the video to accurately depict the objects within the view of the camera. **Figure 4.6** shows the lens distortion removed from the frame. **Figure 4.7** shows an overlay between Faro scan data and the video image. This overlay demonstrates that the distortion was removed from the video image.

Figure 4.4 Frame from raw video from the forward-facing camera.

Figure 4.5 The same frame of video with aspect ratio corrected.

Figure 4.6 The same frame of video with aspect ratio corrected and lens distortion removed.

Figure 4.7 Scan data overlaid on the same frame of video with aspect ratio corrected and lens distortion removed.

Footage from the forward-facing camera on the bus was analyzed to determine positions and speeds for the bus in the moments leading up to the collision. For this analysis, we were provided with an executable video file and an AVI video file. The executable file contained all seven video streams from the bus. The AVI file was a single video stream, the forward-facing view, that had been exported from the executable file data. Double-clicking on the executable file opened the footage in an application called Clip Player. This application displayed a timestamp for the video frames in the format hh:mm:ss. The provided footage began with a timestamp of 19:29:59 and ended with a timestamp of 20:00:00. Thus, the footage encompassed a timeframe of 30 min and 1 sec.

To confirm the frame rate for the forward-facing footage, frames were exported from the Clip Player and saved as still images. Frames were captured starting at timestamp 19:33:12 and ending at timestamp 19:34:04 for a total duration of approximately 52 sec. This timeframe contained 445 unique frames. A duration of 52 sec with 445 unique frames implies an average frame rate of 8.56 fps, within the range of frame rates we obtained based on the duration of the yellow traffic signals. These frames show the bus traveling northbound in the far right lane. They show a northbound car located in the northbound left turn pocket and the pedestrian walking approximately centered in the crosswalk.

We used the camera matching method to track the movement of the bus and the pedestrian across a series of frames leading up to the collision. This analysis utilized distortion-corrected video frames and the known location of the camera within the bus (documented during the inspection of the bus). The images in **Figure 4.8** show sample camera-matched frames. The images on the left are the distortion-corrected video frames, and those on the right are the same video frames with the Faro scan data overlaid onto the image. The camera-matched frames yielded the motion of the camera and therefore of the bus. Some of the matched frames also showed the pedestrian, and we tracked her position in these frames.

Figure 4.8 Sample camera-matched frames.

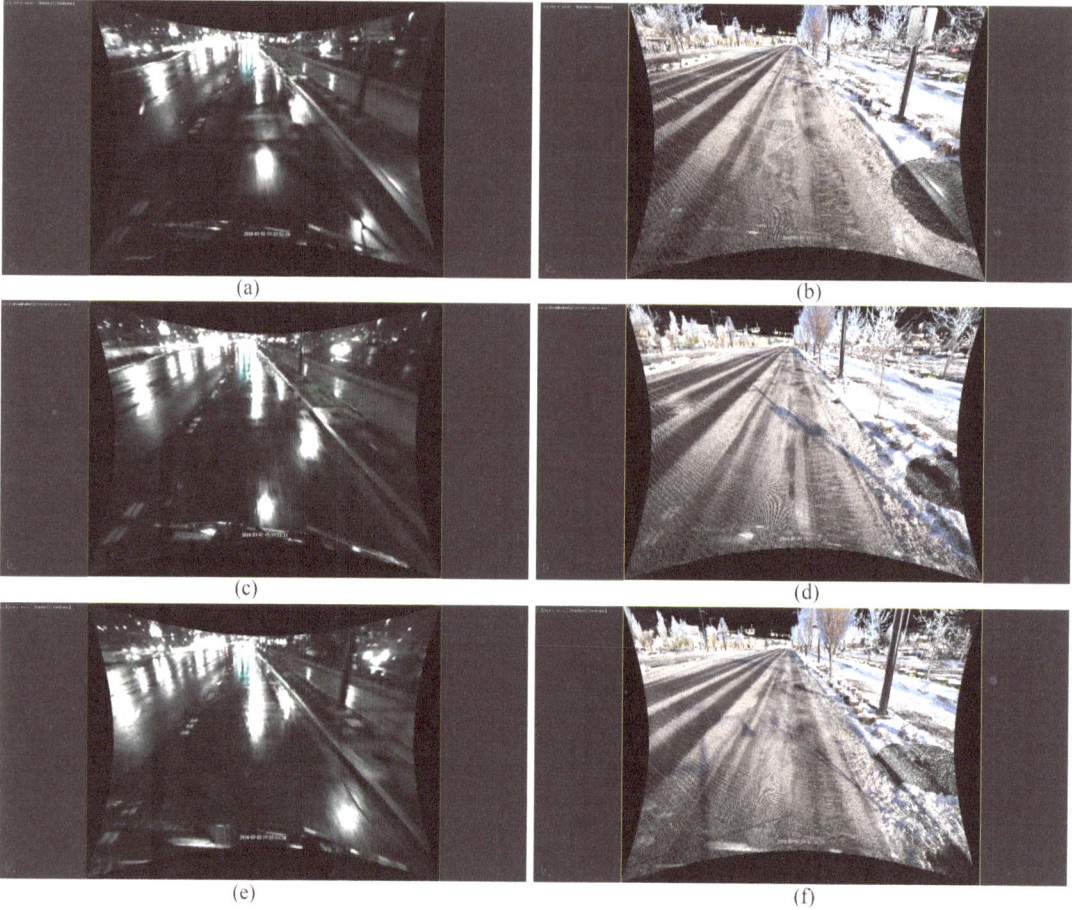

(a)

(b)

(c)

(d)

(e)

(f)

Based on the bus positions obtained from this analysis and the known average frame rate, we calculated that the bus was initially traveling at approximately 34 mph and decelerating as it approached the intersection. The bus slowed to approximately 29 mph, and then, the driver applied the brakes to an emergency level just prior to the impact. The speed at the time of the collision was approximately 28 mph. Pedestrian speeds were also calculated based on the positions obtained from the video tracking. These calculations showed that the pedestrian was traveling between 5.1 and 5.6 ft/s during the final second leading up to impact.

Case Study: Camera Matching to Track Vehicle and Pedestrian Motion

This case study covers the reconstruction of a daytime pedestrian collision that occurred in a store parking lot. The collision involved a Cadillac SRX SUV and occurred when the driver backed out of a parking spot in the store parking lot and struck the pedestrian. The weather at the time was clear and dry. Several surveillance cameras at the store captured the vehicle and pedestrian movements prior to and during the collision. We were provided with six video files from these cameras. The first showed the pedestrian inside the store, checking out at the register. The next showed the pedestrian inside the store, walking from the register to the doors. The third showed the pedestrian walking through the doors of the store. The fourth showed the pedestrian walking out the doors of the store, across the sidewalk, and into the parking lot. The collision itself, along with the motion of the pedestrian and the Cadillac leading up to and following the collision, was captured in the fifth and sixth videos. Sample frames from these views are included in **Figure 4.9**. These frames show the time immediately preceding the collision. The Cadillac SUV is in

the midst of pulling out of the spot, and the pedestrian is walking down the aisle near the Cadillac. These surveillance videos enabled us to determine where the Cadillac was parked prior to the collision and the timing of this vehicle's motion in relation to the pedestrian.

The parking lot where this collision occurred was inspected and digitally mapped using a Faro-brand laser scanner. We were able to arrange to inspect the subject Cadillac at the site—parked in the same spot it was parked prior to the subject collision—and we were able to view the video feed at the store (**Figure 4.10**). The Cadillac SRX was equipped with ultrasonic parking assist and a rear vision camera. According to the owner's manual, the ultrasonic parking assist would operate at speeds less than 5 mph and detect objects up to 8 ft behind the vehicle. However, the manual also stated that this system could not detect pedestrians. When the Cadillac was on and shifted into reverse, there were backup lights that would illuminate, and the camera would turn on.

During inspection of the Cadillac, we documented the screen for the backup camera. When the vehicle was not in reverse, this screen was in a retracted position. When the vehicle was shifted into reverse, the screen would come out of its retracted position and turn on. We took the video of the vehicle being shifted into reverse and the screen coming up and turning on. This video was taken with a frame rate of 24 fps. A sampling of frames from this video is included in **Figure 4.11**. We found that from the time the vehicle was shifted into reverse until the time the screen display turned on was approximately 2.6 sec. The time from when the vehicle was shifted into reverse until the time the screen was fully extended was approximately 3.0 sec.

Figure 4.9 Sample frames from the footage showing the parking lot.

(a)

(b)

Figure 4.10 Subject Cadillac in parking lot and video feed.

(a)

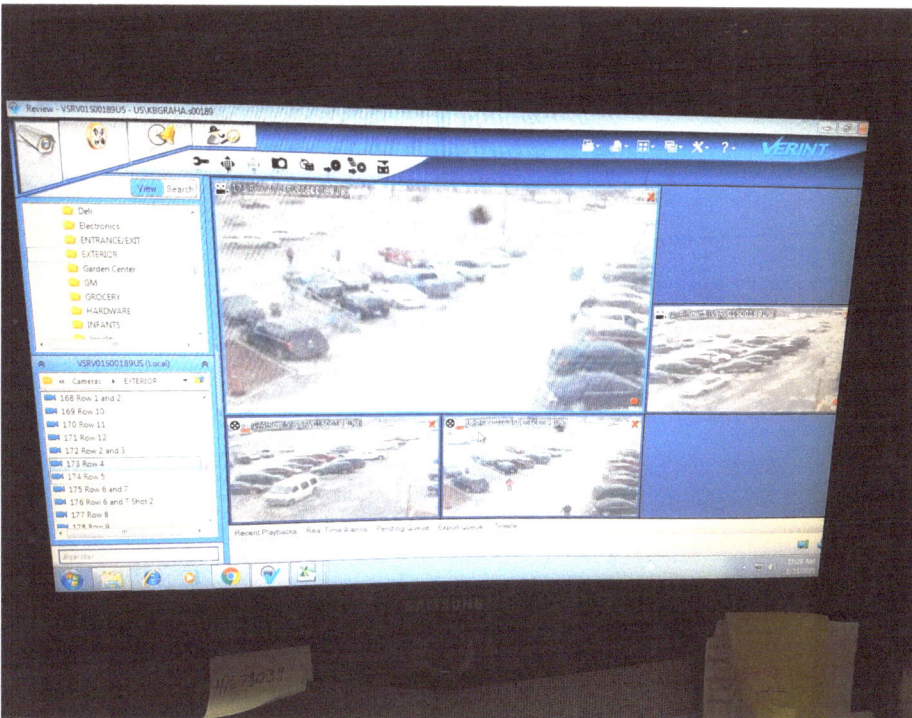

(b)

Figure 4.11 Video frames showing the backup camera activating when the vehicle was shifted into reverse.

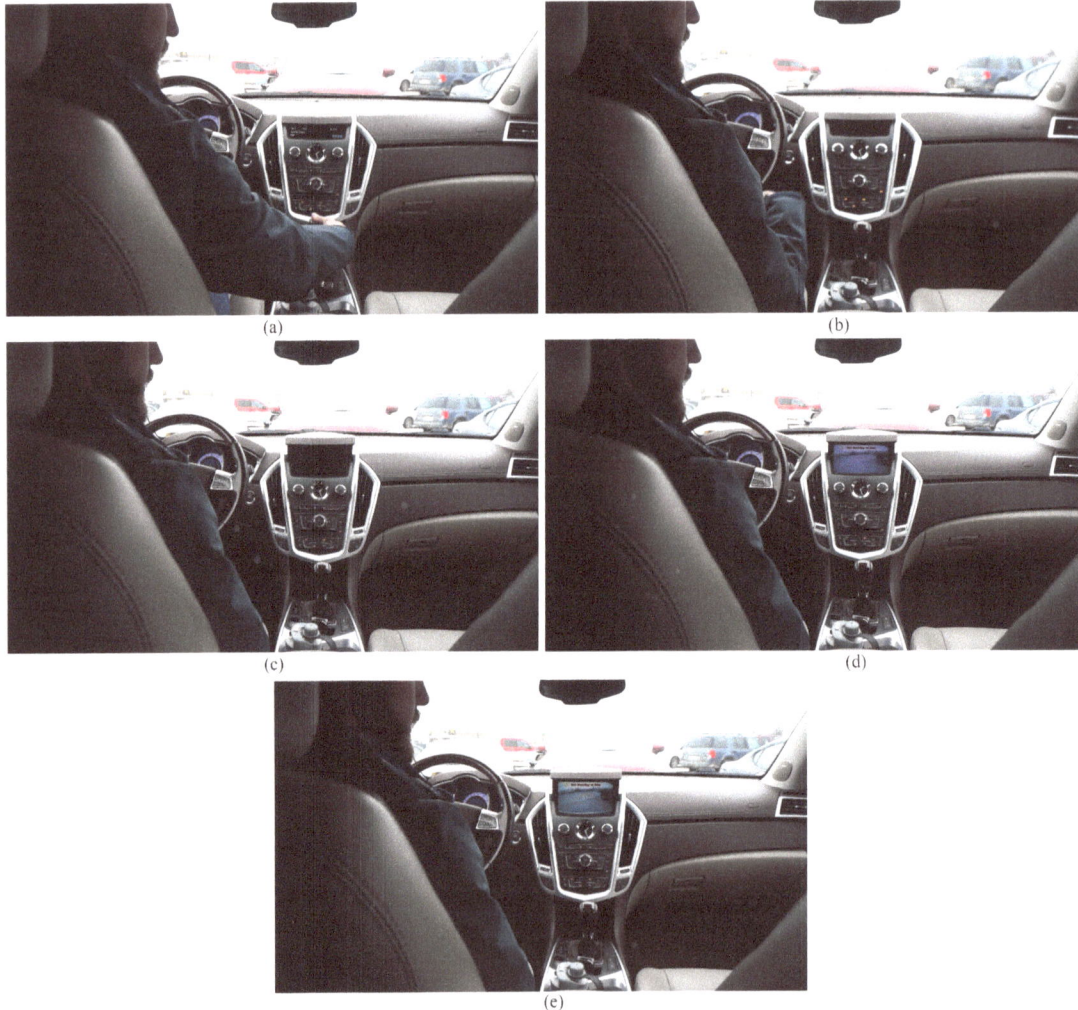

In addition to the video of the backup camera screen coming out of its retracted position, we also took the video of the view from the backup camera. A surrogate pedestrian walked a route similar to that walked by the subject pedestrian, and the view provided by the screen was recorded. Sample frames from this video are included in **Figure 4.12**. In addition to inspecting the subject Cadillac at the site, an exemplar Cadillac SRX was also inspected. During this inspection, we mapped the view from the interior rear-view mirror, the two side mirrors, and the backup camera. We mapped out the area visible in each mirror or camera using cones and then used a Faro laser scanner to capture measurements of these cones. An image showing our scan data is included in **Figure 4.13**. This image shows a cone identifying the area visible in each mirror or camera. The blue area is visible with the backup camera, the green areas are visible in the side mirrors, and the red area is visible in the interior rear-view mirror.

Figure 4.12 Video frames showing the view with the backup camera.

Figure 4.13 Areas visible with mirrors and backup camera.

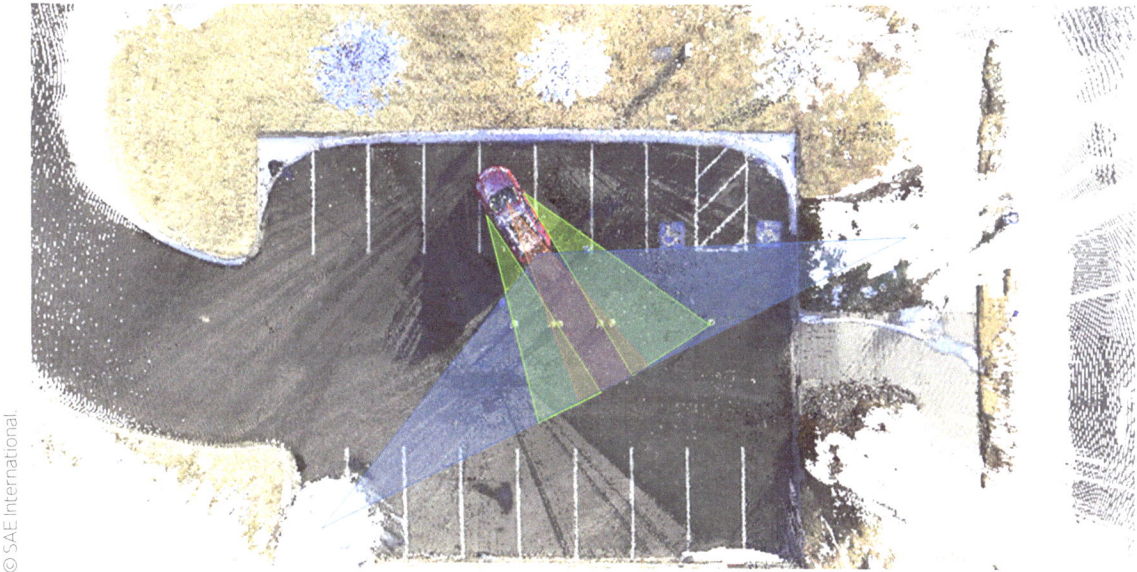

Figure 4.14 Tracking the motion of the Cadillac and the pedestrian after achieving alignment between the scene scan data and the depicted scene.

The videos of the subject collision were determined to have a frame rate of 5 fps. We extracted these frames from the video and analyzed them frame by frame using the camera matching method to determine the relative motion of the Cadillac and the pedestrian. This process began with removing lens distortion from the video frames. The images in **Figure 4.14** show the results of the tracking. Each image shows a single video frame in which we have used the camera matching technique to locate the Cadillac and the pedestrian. We also located three additional vehicles parked near the Cadillac.

Based on the positions obtained from the video tracking and on the known frame rate, speeds were calculated for the Cadillac and the pedestrian. This

resulted in the conclusion that the pedestrian was walking at a speed of approximately 3.9 ft/s. The graph in **Figure 4.15** shows the calculated speeds for the Cadillac. As this graph shows, the driver of the Cadillac accelerated her vehicle to a speed of 4.1 mph (6.0 ft/s) before applying the brakes and stopping. Once the relative movement of the Cadillac and the pedestrian was known, we concluded that, had the driver of the Cadillac waited for the display of her backup camera to turn on prior to backing, the pedestrian would have been visible in the view of the camera. **Figure 4.16** is a graphic showing this. In this graphic, the pedestrian is shown with a blue circle. The shaded areas show the areas visible with each mirror and camera. The blue shading is the area visible with the camera.

Figure 4.15 Cadillac speed determined from tracked positions.

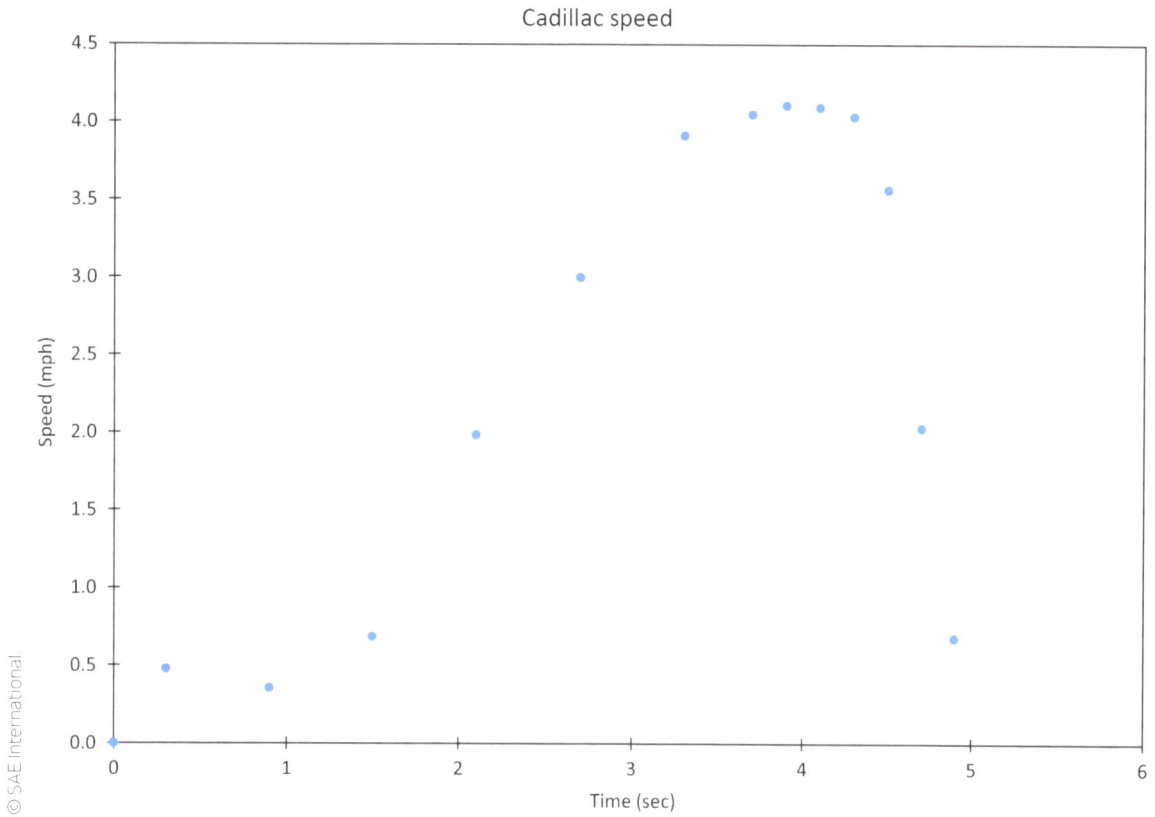

Figure 4.16 Graphic showing pedestrian within viewing area of backup camera.

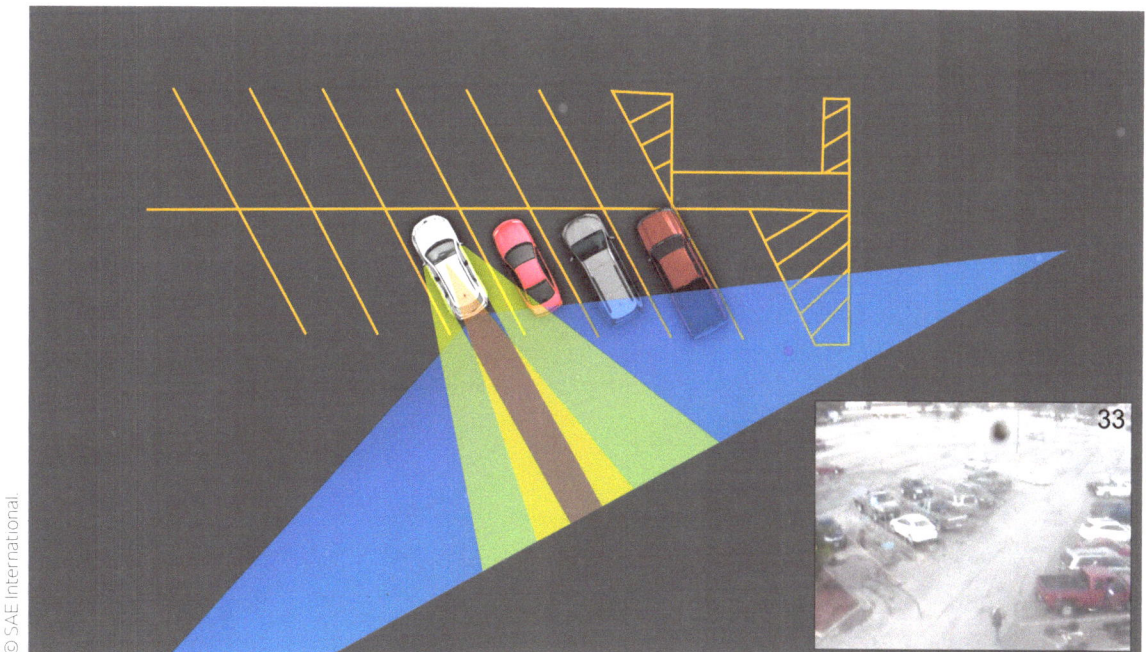

Sources for Site Geometry

Regardless of the method utilized for tracking the vehicle motion in a video, the process requires information about the geometry of the environment depicted in the video. There are several available options for obtaining this site information, including mapping of the site with a laser scanner, mapping of the site using image-based scanning, developing a scene using publicly available lidar data with aerial imagery, or aerial imagery alone. Terpstra et al. [4.6] noted that "in instances where there have been significant site changes, where there is limited or no site access, and where budgeting and timing constraints exist, a three-dimensional environment can be created using publicly available imagery and aerial LiDAR." The USGS has developed a database of lidar data for a significant portion of the US. These data, which are referred to as the 3DEP, are publicly available and downloadable from the USGS website. The program was formed in 2012, and the data became available to the public in 2015. Data are not yet available everywhere in the US, but the coverage continues to grow. A current coverage map is accessible on the USGS website. Several peer-reviewed publications have validated the use of these data for accident reconstruction [4.6, 4.7, 4.8, 4.9].

The characteristics of these data are reported in terms of resolution and accuracy. The ASPRS has established accuracy standards for classifying lidar elevation data [4.10], and the 3DEP has adopted the quality standards for resolution and accuracy listed in **Table 4.1**.[1] Terpstra et al. [4.8] define the accuracy as "the closeness of an estimated value (measured or computed, for example) to a standard or accepted value of a particular quantity together with an explicit reference to the specific standard or accepted value."

Table 4.1 Lidar quality level requirements for the 3DEP.

Quality level	Vertical accuracy, RMSEz, cm (in.)	Nominal pulse spacing, m (ft)
QL0	5 (2.0)	≤0.35 (1.1)
QL1	10 (3.9)	≤0.35 (1.1)
QL2	10 (3.9)	≤0.71 (2.3)
QL3	20 (7.9)	≤1.41 (4.6)

© SAE International

Determining the Frame Rate of a Video

To obtain speeds from tracked vehicle positions, the reconstructionist will need to determine the frame rate of the video and, from this, the time interval that elapsed between the tracked positions. The time that elapsed between the capture of two successive frames in a video is commonly referred to as the frame interval. The frame rate, on the other hand, is the number of individual frames that are captured during each second of the footage. Sometimes the terms

1 https://www.usgs.gov/3d-elevation-program/topographic-data-quality-levels-qls

"frame interval" and "frame rate" are used interchangeably. Examination of the metadata of a video file can be carried out to determine the frame rate and frame interval. There are multiple software options for examining a video file, including the following:

- Operating File System
- FFMPEG
- Axon Investigate (formerly INPUT-ACE)
- Amped Five
- Exiftool
- MediaInfo
- Adobe Premiere
- VideoLAN Client

The underlying structure of a media file will consist of one or more data streams (video, audio, telemetry, etc.) and metadata. These components are saved within the container or wrapper. The metadata typically include information related to how the stream data should be played back by specifying the codec, frame rate, pixel matrix, color space, and bit rate, among other things. In other words, the individual frame data and the protocol for playing back the video are the essence of the file, while the container is the carrier within which these data are stored. We could give the metaphor of a suitcase representing the container, and the contents of the suitcase—an audio tape, stack of photos (video stream), and a three-ring work instructions binder (codec, frame rate, etc.)—being the data within the container. The container can be changed (different suitcase brands, sizes, shapes, etc.) without altering the contents.

An example of the metadata stored by a video file is included next. The metadata revealed using MediaInfo (**Figure 4.17**) list data specific to the file, including the average frame rate, encode date, Group of Pictures (GOP) structure, aspect ratio, and resolution. When frames of the same video are played back within Axon Investigate (**Figure 4.18**), the display includes metadata within the display window, including the frame type, total time, position, and a display of the audio stream wave. FFMPEG can be used to delve deeper into the media by producing an XML of the frame analysis, which can be opened and analyzed using Excel. Frame analysis goes deeper and shows the individual frame timing, if available, and sometimes will specify the frame type (I, P, or B) and the frame size and other pertinent details of the file.

Figure 4.17 Metadata obtained using MediaInfo.

■)) MediaArea.net/MediaInfo - C:\Users\csmith\EEC Dropbox\Connor Smith\Pedestrian Crash Reconstruction CLE\Metadata Example.mp4

File ⋮ View ⋮ Options ⋮ Debug ⋮ Help ⋮ Language ⋮ Winxvideo AI-Enhance/Convert Videos (AD)

General

Complete name :	C:\Users\csmith\EEC Dropbox\Connor Smith\Pedestrian Crash Reconstruction CLE\Metadata Example.mp4
Format :	MPEG-4
Format profile :	Base Media / Version 2
Codec ID :	mp42 (mp42/mp41)
File size :	14.4 MiB
Duration :	6 s 73 ms
Overall bit rate mode :	Variable
Overall bit rate :	19.9 Mb/s
Frame rate :	29.970 FPS
Encoded date :	2024-07-12 18:30:50 UTC
Tagged date :	2024-07-12 18:30:53 UTC
TIM :	00:00:00:00
TSC :	30000
TSZ :	1001

Video

ID :	1
Format :	AVC
Format Info :	Advanced Video Codec
Format profile :	Main@L4.1
Format settings :	CABAC / 4 Ref Frames
Format settings, CABAC :	Yes
Format settings, Reference frames :	4 frames
Codec ID :	avc1
Codec ID Info :	Advanced Video Coding
Duration :	6 s 73 ms
Bit rate :	19.5 Mb/s
Width :	1 920 pixels
Height :	1 080 pixels
Display aspect ratio :	16:9
Frame rate mode :	Constant
Frame rate :	29.970 (30000/1001) FPS
Color space :	YUV
Chroma subsampling :	4:2:0
Bit depth :	8 bits
Scan type :	Progressive
Bits/(Pixel*Frame) :	0.313
Stream size :	14.1 MiB (98%)
Writing library :	AVC Coding
Language :	English
Encoded date :	2024-07-12 18:30:51 UTC
Tagged date :	2024-07-12 18:30:51 UTC
Color range :	Limited
Color primaries :	BT.709
Transfer characteristics :	BT.709
Matrix coefficients :	BT.709
Codec configuration box :	avcC

Audio

ID :	2
Format :	AAC LC
Format Info :	Advanced Audio Codec Low Complexity
Codec ID :	mp4a-40-2
Duration :	6 s 73 ms
Source duration :	6 s 101 ms
Bit rate mode :	Variable
Bit rate :	317 kb/s
Maximum bit rate :	392 kb/s
Channel(s) :	2 channels
Channel layout :	L R
Sampling rate :	48.0 kHz
Frame rate :	46.875 FPS (1024 SPF)
Compression mode :	Lossy
Stream size :	235 KiB (2%)
Source stream size :	236 KiB (2%)
Language :	English
Encoded date :	2024-07-12 18:30:51 UTC
Tagged date :	2024-07-12 18:30:51 UTC

(Continued)

Figure 4.17 (Continued) Metadata obtained using MediaInfo.

Type	Stream	Frame #	Key frame	Pts	Pts (seconds)	Best effort (seconds)	Duration (seconds)	Position	Size	Width	Height	Pixel Format	Sample Aspect Ratio	Frame Type	Coded Picture Number	Display Picture Number
video	1	0	TRUE	0	0	0	0.033367	61496	377060	1920	1080	yuv420p	0:1	I	0	0
video	1	1	FALSE	1001	0.033367	0.033367	0.033367	454549	11109	1920	1080	yuv420p	0:1	B	2	0
video	1	2	FALSE	2002	0.066733	0.066733	0.033367	438556	15993	1920	1080	yuv420p	0:1	P	1	0
video	1	3	FALSE	3003	0.1001	0.1001	0.033367	486846	37068	1920	1080	yuv420p	0:1	B	4	0
video	1	4	FALSE	4004	0.133467	0.133467	0.033367	465658	21188	1920	1080	yuv420p	0:1	P	3	0
video	1	5	FALSE	5005	0.166833	0.166833	0.033367	547561	62944	1920	1080	yuv420p	0:1	B	6	0
video	1	6	FALSE	6006	0.2002	0.2002	0.033367	523914	23647	1920	1080	yuv420p	0:1	P	5	0
video	1	7	FALSE	7007	0.233567	0.233567	0.033367	653812	78602	1920	1080	yuv420p	0:1	B	8	0
video	1	8	FALSE	8008	0.266933	0.266933	0.033367	610505	43307	1920	1080	yuv420p	0:1	P	7	0
video	1	9	FALSE	9009	0.3003	0.3003	0.033367	800989	91832	1920	1080	yuv420p	0:1	B	10	0
video	1	10	FALSE	10010	0.333667	0.333667	0.033367	732414	55129	1920	1080	yuv420p	0:1	P	9	0
video	1	11	FALSE	11011	0.367033	0.367033	0.033367	966006	75016	1920	1080	yuv420p	0:1	B	12	0
video	1	12	FALSE	12012	0.4004	0.4004	0.033367	892821	73185	1920	1080	yuv420p	0:1	P	11	0
video	1	13	FALSE	13013	0.433767	0.433767	0.033367	1108027	31324	1920	1080	yuv420p	0:1	B	14	0
video	1	14	FALSE	14014	0.467133	0.467133	0.033367	1041022	67005	1920	1080	yuv420p	0:1	P	13	0
video	1	15	FALSE	15015	0.5005	0.5005	0.033367	1244889	25326	1920	1080	yuv420p	0:1	B	16	0
video	1	16	FALSE	16016	0.533867	0.533867	0.033367	1139351	105538	1920	1080	yuv420p	0:1	P	15	0
video	1	17	FALSE	17017	0.567233	0.567233	0.033367	1373271	12707	1920	1080	yuv420p	0:1	B	18	0
video	1	18	FALSE	18018	0.6006	0.6006	0.033367	1270215	103056	1920	1080	yuv420p	0:1	P	17	0
video	1	19	FALSE	19019	0.633967	0.633967	0.033367	1533482	14805	1920	1080	yuv420p	0:1	B	20	0
video	1	20	FALSE	20020	0.667333	0.667333	0.033367	1385978	133901	1920	1080	yuv420p	0:1	P	19	0
video	1	21	FALSE	21021	0.7007	0.7007	0.033367	1650593	16906	1920	1080	yuv420p	0:1	B	22	0
video	1	22	FALSE	22022	0.734067	0.734067	0.033367	1548287	102306	1920	1080	yuv420p	0:1	P	21	0
video	1	23	FALSE	23023	0.767433	0.767433	0.033367	1865990	19540	1920	1080	yuv420p	0:1	B	24	0
video	1	24	FALSE	24024	0.8008	0.8008	0.033367	1667499	198491	1920	1080	yuv420p	0:1	P	23	0
video	1	25	FALSE	25025	0.834167	0.834167	0.033367	2077876	21725	1920	1080	yuv420p	0:1	B	26	0
video	1	26	FALSE	26026	0.867533	0.867533	0.033367	1885530	192346	1920	1080	yuv420p	0:1	P	25	0
video	1	27	FALSE	27027	0.9009	0.9009	0.033367	2237087	29319	1920	1080	yuv420p	0:1	B	28	0
video	1	28	FALSE	28028	0.934267	0.934267	0.033367	2099601	137486	1920	1080	yuv420p	0:1	P	27	0
video	1	29	FALSE	29029	0.967633	0.967633	0.033367	2418280	36172	1920	1080	yuv420p	0:1	B	30	0
video	1	30	FALSE	30030	1.001	1.001	0.033367	2266406	139253	1920	1080	yuv420p	0:1	P	29	0

Figure 4.18 Information about a single frame of video being viewed in Axon Investigate.

Note that the frame rate reported in the metadata of a video file may not be the rate of the original footage but could be a rate specified by the codec for playback of that version of the file. In some cases, the integrity of the video file has been upheld, and the file has not been changed, altered, or corrupted from its original state. Sometimes, though, the original file may have been altered by the time it reaches an accident reconstructionist or video analyst. Sometimes, a video file is converted from one format to another, and the video file may end up being transcoded. When a video is recorded by a camera, the data are encoded and stored as a video stream. During playback, this data stream is decoded and presented within the software. If the file is saved into another format, redetected for time or cropped for size, or modified to play back at a different frame rate than the original recording, the data are re-encoded into a new video stream with the updated video file. The metadata of this version of the file reflect the most recent codec, rather than the original codec. Thus, the frame rate listed in the metadata should be checked with other methods, if possible.

After reviewing the metadata, the next step in a typical video analysis workflow is reviewing the individual frames. For this process, the individual frames from the video stream are exported and reviewed to verify the metadata and to identify any anomalies or differences from the properties reported in the metadata. Verification of the resolution and aspect ratio can be completed using an individual frame. Borders around videos, black bars for titles at the top or bottom of videos, and overlayed or embedded timestamps can be examined and considered. The aspect ratio can be checked by evaluating the shape of objects with known geometry shown in the video, such as signage and spherical objects.

Potential issues related to the frame rate of a video can be identified by paging through the individual frames in sequence. This can reveal repeated frames, dropped frames, and larger-than-expected jumps in the position of objects or vehicles. Dropped and repeated frames are sometimes found in videos that have been transcoded between different frame rates. Pixel tracking on multiple successive frames is a common method for investigating whether a frame rate is constant or variable. Pixel tracking is done by tracking an object that passes through a video in successive frames and overlaying that track onto one frame to analyze the pattern it produces. This process utilizes objects that are moving at a relatively constant speed and are traveling across the frame rather than into or out of the frame. **Figure 4.19** shows two examples of pixel tracking. Pixel tracking on these images utilizes the same point on the center of a tire driving through the frame. Pixel tracking on the left shows the results when applied to a video with a constant frame rate. While the distance between each pixel is not constant, the size is constantly increasing and is indicative of a constant frame rate. Pixel tracking on the right shows the results of pixel tracking on a variable frame rate video. In this image, the distance between the pixels for successive frames is not constant or varying regularly.

Figure 4.19 Pixel tracking to assess the regularity of the frame rate.

(a)

(b)

Video Compression and Frame Types

Video compression eliminates redundant information in a video file to reduce the size of the file and thus to reduce the data storage and playback requirements. Compression is carried out in two forms: spatial compression and temporal compression [4.7]. Both spatial and temporal compression are lossy compression types, meaning that some of the original video information is lost during the compression process. This is in contrast to lossless compression, in which the original video information is retained. Spatial compression is compression within a video frame that reduces the pixel data within the frame. Each frame is treated independently (intra-frame compression). Spatial compression takes advantage of similarities and patterns between pixels close to each other by eliminating redundant color information, reducing unnecessary detail, and grouping individual pixels into blocks of pixels. As an example, JPEG—often thought of as a file format—is actually a lossy spatial compression methodology developed by the Joint Photographic Experts Group. The JPEG compression methodology uses the following procedure:

- Color transformation (RGB to YCbCr)
- Chroma subsampling
- Discrete cosine transformation (DCT)
- Quantization
- Entropy encoding

The first step in this process is transforming the colors from the red, green, and blue (RGB) color space to a model based on luminance and chrominance (YCbCr). In the YCbCr color space, "Y" represents the luminance (brightness), while "Cb" and "Cr" represent the blue-difference and red-difference chrominance components. The human eye is more sensitive to luminance than to color, so converting the RGB color space to the YCbCr color space allows the color value to be subsampled with a less perceptible effect to the overall image than if the RGB values were subsampled. The luminance values are next converted to the frequency domain and are represented with a single 8×8 pixel block using DCT. Quantization is the final step prior to encoding, which lowers the precision of the data stored in the DCT coefficient matrix.

The magnitude of spatial compression will be determined by how similar the brightness and colors need to be in order to be reduced to the same color. The wider the boundaries around what it means for two colors to be similar, the higher the level of compression. Spatial compression also takes advantage of similarities and patterns within a frame during DCT and quantization. The amount of compression can be specified by the amount of quantization and the chroma subsampling. For individual frames in a video, the magnitude of compression can be controlled by the byte target of each frame and the overall bitrate of the camera system. The images in **Figure 4.20** show the effect of higher levels of spatial compression on an image, progressing from the least compression on the left to the highest on the right.

Pedestrian Accident Reconstruction **159**

Figure 4.20 Increasing spatial compression left to right.

Spatial compression reduces the color information that is stored for any individual frame. While spatial compression could have a detrimental impact for certain types of analysis, in the context of collision reconstruction, color information is not typically the focus of the analysis. Instead, the focus is on the motion of objects, and for this reason, the collision reconstructionist will usually be more concerned with temporal compression. That said, spatial compression could influence how the edges of a vehicle or other object moving through the view of the camera are depicted. Thus, if the moving object fills only a small part of the frame, spatial compression could result in uncertainty related to the position of the object within any particular frame. That uncertainty may need to be quantified, and its significance evaluated. This will not be an issue, though, as long as the object fills an adequate portion of the frame for the analysis to not depend on the precise location of the edges of the object. Furthermore, if frames prior to and after the frame being analyzed can be utilized as bounds or restrictions on the location of the object in the frame, then the uncertainty in position due to spatial compression can be reduced.

Temporal compression eliminates redundant information across a series of frames (interframe compression). It utilizes the similarity in data between frames that are close in time. For example, for a fixed-view surveillance camera with a vehicle moving through the view of the camera, much of the background behind the moving vehicle will be unchanged from frame to frame. The pixels representing this background do not need to be stored for each individual frame, since this information can be borrowed from the surrounding frames where it is stored. Temporal compression uses the following frame types:

- I-frame (intra-coded frame)—This is a frame in which all of the pixels in the frame are newly encoded information. There is no temporal compression in an I-frame. These are key frames from which P and B frames can borrow information.

- P-frame (predicted frame)—This is a frame in which only changes from the prior frame are stored.

- B-frame (bidirectional predicted frame)—This is a frame in which the information storage is further reduced by looking not just to the prior I- or P-frame but also to the I- or P-frame ahead.

The GOP of a video designates the structure of I-, P-, and B-frames. The GOP is quantified by examining the location of I-frames and the pattern of P- and B-frames, with one GOP typically defined by the series of frames that start with an I-frame and ends just prior to the next I-frame. When a video is encoded, the algorithm examines P- and B-frames in context with the frames around it to determine what information is new and then determines whether a previous block of pixels can be used as a prediction for the new information or whether the entire block or sub-block should be encoded. Crouch and Cash [4.11]: "What this means is that pixels, and therefore objects, shown in P and B-frames can in effect be a *prediction* of their true position, particularly when pixels have been moved/copied." [Emphasis in original.] It is important to note, though, that in all frame types, there can be newly encoded information. The threshold of change that the algorithm looks for determines how much change needs to occur to encode new information, and B-frames have a lower threshold than P-frames. The appropriateness of using B- and P-frames for analysis can be evaluated on a case-by-case basis based on what newly encoded information is present in the frames and on how much of the frame the tracked object fills.

The use of temporal compression can be visualized by running a macroblock analysis on a video. A macroblock analysis provides the analyst with encoding information regarding each block or sub-block of pixels. Forensic video analysis software can be utilized for performing this analysis, and the result is the frames of a video presented with a color overlay representing the temporal process used. A key or legend is provided with the analysis to interpret block colors and any block motion vectors present. The series of frames in **Figure 4.21** show the results from a macroblock analysis. The first frame in the sequence is an I-frame, and the entire frame consists of newly encoded data. In this frame, the purple overlay designates full macroblocks coded with intra-prediction (16 × 16), and the pink overlay designates macroblocks divided into 16 smaller blocks and coded with intra-prediction (4 × 4). The second and third frames are P-frames, with the colored blocks representing blocks referenced from a past frame (green), i.e., predicted, and blocks representing newly encoded information (pink and purple).

Figure 4.21 Example of macroblock analysis.

References

4.1.   Neale, W., Hessel, D., and Terpstra, T., "Photogrammetric Measurement Error Associated with Lens Distortion," SAE Technical Paper 2011-01-0286 (2011), doi:https://doi.org/10.4271/2011-01-0286.

4.2.   Terpstra, T., Miller, S., and Hashemian, A., "An Evaluation of Two Methodologies for Lens Distortion Removal When EXIF Data Is Unavailable," SAE Technical Paper 2017-01-1422 (2017), doi:https://doi.org/10.4271/2017-01-1422.

4.3.   Beauchamp, G., Pentecost, D., Koch, D., Hashemian, A. et al., "Speed Analysis from Video: A Method for Determining a Range in the Calculations," SAE Technical Paper 2021-01-0887 (2021), doi:https://doi.org/10.4271/2021-01-0887.

4.4.   Terpstra, T., Hashemian, A., Gillihan, R., King, E. et al., "Accuracies in Single Image Camera Matching Photogrammetry," SAE Technical Paper 2021-01-0888 (2021), doi:https://doi.org/10.4271/2021-01-0888.

4.5.   Simacek, D., Tovar, J., Famiglietti, N., Shkolkin, V. et al., "Stationary and Moving Camera Video Analysis Compared to Known Reference System," SAE Technical Paper 2021-01-0879 (2021), doi:https://doi.org/10.4271/2021-01-0879.

4.6.   Terpstra, T., McDonough, S., Helms, E., Beier, S. et al., "Video and Object Tracking for Speed Determination Using Aerial LiDAR," SAE Technical Paper 2024-01-2483 (2024), doi:https://doi.org/10.4271/2024-01-2483.

4.7.   Terpstra, T., Dickinson, J., and Hashemian, A., "Using Multiple Photographs and USGS LiDAR to Improve Photogrammetric Accuracy," *SAE Int. J. Trans. Safety* 6, no. 3 (2018): 193-216, doi:https://doi.org/10.4271/2018-01-0516.

4.8.   Terpstra, T., Dickinson, J., Hashemian, A., and Fenton, S., "Reconstruction of 3D Accident Sites Using USGS LiDAR, Aerial Images, and Photogrammetry," SAE Technical Paper 2019-01-0423 (2019), doi:https://doi.org/10.4271/2019-01-0423.

4.9.   Terpstra, T., Mckelvey, N., King, E., Hashemian, A. et al., "Aerial Photoscanning with Ground Control Points from USGS LiDAR," *SAE Int. J. Adv. & Curr. Prac. in Mobility* 4, no. 4 (2022): 1384-1393, doi:https://doi.org/10.4271/2022-01-0833.

4.10. American Society for Photogrammetry and Remote Sensing (ASPRS), "ASPRS Positional Accuracy Standards for Digital Geospatial Data," Edition 2, Version 1.0, August 23, 2023.

4.11. Crouch, M. and Cash, S., *Video Analysis in Collision Reconstruction*, 2nd ed. (South Croydon, England: Forensic Collision Investigation & Reconstruction LTD, 2023).

Chapter 05

Event Data Recorders in Pedestrian Accident Reconstruction

EDR Events and Pedestrian Collisions

EDRs have led to a massive increase in the evidence available to accident reconstructionists. These systems have opened access to the phase of the crash where drivers are perceiving and responding to a hazard (or not), prior to the point where there is any discernible physical evidence on the road. Thirty years ago, accident reconstructionists had little, if any, access to this human factors phase of the collision. Of course, there are still many instances where a vehicle and its EDR data are not preserved. Even in these instances, though, the reconstructionist is working in a more data-rich environment than their pre-EDR predecessors, since the abundance of EDR data has led reconstructionists to a deeper understanding and appreciation of the human factors side of the crashes they analyze.

The Crash Data Retrieval (CDR) system, which is the commercially available system that enables access to the EDR data for many vehicles, was released in 2000 [5.1]. Thus, reconstructionists are about 25 years into having access to at least some sensor-measured data from vehicles for use in their work. The CDR system gave reconstructionists access to crash data for General Motors (GM) vehicles as far back as model year 1994 and to pre-crash data for GM vehicles as far back as model year 1999. Ford vehicles became supported in 2003 (for vehicles as old as model year 2001), and Chrysler vehicles became supported in 2007 (for vehicles as old as model year 2005). Since then, many other manufacturers have been added (Toyota/Lexus, Volkswagen, Mazda, Subaru, Infiniti, Nissan, Honda/Acura, and others). EDR data from Hyundai and Kia vehicles are also available, but accessing these data requires a tool manufactured by Global Information Technologies (GIT).

The specific data likely to be present on an EDR from a particular vehicle can be checked in a database that is provided along with the CDR software. In some instances, it can also be helpful to examine exemplar CDR reports from similar make and model vehicles involved in other crashes. Examination of exemplar CDR reports can often be carried out through the National Highway Traffic Safety Administration's (NHTSA's) Crash Investigation Sampling System (CISS).[1] According to NHTSA's website,[2] "CISS collects detailed crash data to help scientists and engineers analyze motor vehicle crashes and injuries. CISS collects data on a representative sample of minor, serious, and fatal crashes involving at least one passenger vehicle – cars, light trucks, SUVs, and vans – towed from the scene. After a crash has been sampled, trained Crash Technicians obtain data from crash sites by documenting scene evidence such as skid marks, fluid spills, and struck objects. They locate the vehicles involved, document the crash damage, and identify interior components that were contacted by the occupants. On-site inspections are followed up with confidential interviews of the crash victims and a review of medical records for injuries sustained in the crash. CISS uses emerging technologies and methods to acquire quality data." Often, the data collected by CISS investigators includes event data from the involved ACMs of vehicles.

When it comes to pedestrian collisions, there is no certainty that the collision will be captured by the EDR, particularly the EDRs contained in passenger vehicle ACMs. Suppose, for example, that a 3000-lb (1361 kg) sedan impacts a 200-lb (90.7 kg) pedestrian at a speed of 40 mph (64.4 km/h), accelerating the pedestrian to a speed of 28 mph (45.1 km/h) over a time period of 150 ms (a projection efficiency of 0.7). If there is no braking by the driver of the sedan during the 150-ms collision, then the sedan will experience a change in velocity (ΔV) of about 1.9 mph (3.1 km/h). For most modern passenger vehicle EDRs, ΔVs of this magnitude will not be high enough to trigger the recording of an event. If we assume that the driver of the sedan is braking severely during the collision, generating a deceleration of 1g, then during the 150-ms collision, the vehicle will experience a ΔV of about 3.3 mph (5.3 km/h) from braking. The additional change in velocity from the collision could then take the total ΔV for the 150-ms contact over the 5-mph (8 km/h) threshold to register and record an event.

An assumption inherent in these illustrative numbers, though, is that the ACM records and registers the entire velocity change. However, during a pedestrian collision, the accelerations experienced by the vehicle will usually be low, and most EDRs have an acceleration threshold that must be exceeded before the ΔV will be recorded. The accelerations experienced by the vehicle prior to this threshold being exceeded can be missed by the EDR, and this will lead to the ACM-recorded ΔV being under-reported. Prior studies have indicated that the triggering threshold for passenger vehicle EDRs is usually in the range of 1 to 2g. This error source in the EDR-reported ΔV can be exacerbated by some EDRs having a built-in positive offset in the accelerometers that can contribute to under-reporting accelerations and ΔVs for frontal impacts. The magnitude of this offset can vary from one vehicle to another and one EDR generation to another [5.2, 5.3, 5.4, 5.5].

Since September 1, 2012, 49 CFR Part 563 in the US has imposed reporting requirements on

---

1   https://crashviewer.nhtsa.dot.gov/CISS/SearchFilter#
2   https://www.nhtsa.gov/crash-data-systems/crash-investigation-sampling-system

passenger car EDRs. This regulation did not require passenger car manufacturers to equip vehicles with EDRs, but for vehicles that are equipped, it created reporting requirements on the data and it required the data to be accessible with publicly available tools. Part 563 has required pre-crash data elements to be reported for 5 sec with a frequency of 2 Hz. Recently, NHTSA has issued an amendment to Part 563 that takes effect from September 1, 2027. This amendment extends the EDR recording period from 5 sec of 2 Hz pre-crash data to 20 sec of 10-Hz data [5.6]. A study by Chen et al. [5.7] had concluded that "the current 5-second EDR recording duration was not sufficient to capture all driver pre-crash maneuver initiations in rear-end, intersection, as well as road departure crashes…EDRs failed to capture driver pre-crash braking initiation in approximately 35% of the rear-end, intersection, and road departure crashes…In addition, current EDR recording duration failed to capture driver pre-crash steering maneuver initiation in 64-88% of all rear-end, intersection, and road departure crashes…Similarly, the analysis…shows that current EDRs failed to capture up to nearly 56% of all driver pre-crash braking initiation in rear-end crashes." This study concluded that "an EDR recording duration of 20 seconds would adequately capture driver pre-crash inputs to the vehicle, i.e. braking, steering, and acceleration, in rear crashes, intersection crashes, and road departures."

In a study published in 2024, Watson et al. [5.8] studied the field performance of passenger vehicle EDRs in terms of the probability that an event would be recorded at the Part-563-required ΔV threshold of 8 km/h (5 mph) occurring within 150 ms. This study utilized EDR-reported and reconstructed ΔV data from 3960 cases contained in the CISS database

maintained by NHTSA. These authors noted that "the EDR trigger threshold set at a ΔV of 8 km/h ensures that EDR data captures the low end of the crash severity spectrum. Vehicles built after the Part 563 compliance date have higher probabilities of capturing low speed events on the EDR. This is an indication that the Part 563 trigger threshold is effective in increasing the amount of EDR data available for accident reconstruction and safety research." These authors found that "vehicles manufactured by Toyota had lower ΔV thresholds compared to other manufacturers. For Toyota manufactured vehicles, the probability of an EDR event at 8 km/h ranged from 87% to 99%. EDR event probabilities for non-Toyota vehicles in vehicle-to-vehicle collisions ranged from 58% to 93% for Part 563 compliant vehicles and 33% to 83% for pre-Part 563 vehicles. The persistence of EDR events in memory was analyzed using the number of ignition cycles present between events and imaging of the EDR, finding average duration of 3595 ignition cycles for pre-existing EDR events unrelated to the CISS case."

A 2004 study by Fugger, Randles, and Eubanks [5.9] reported a series of 37 staged pedestrian collisions that utilized pre-Part 563 EDR-equipped vehicles—a 1997 Chevrolet Venture (24.5:1 weight ratio with the test dummy), a 2001 Chevrolet Cavalier LS (15:1 weight ratio with the test dummy), an Oldsmobile Aero (17.9:1 weight ratio with the test dummy), and a Saturn SC2 (14.5:1 weight ratio with the test dummy). The dummy was an Alderson Research Labs Model CG-95 anthropometric test dummy with a height of 75.5 in. (1.92 m) and a weight of 169 lb (76.8 kg). Vehicle impact speeds between 5.2 and 39 mph were utilized. Of the 37 collisions, 16 resulted in a non-deployment event being recorded on the EDR. The severity of the other 21 collisions was either not sufficient to wake up the module or not sufficient to overwrite an existing non-deployment event on the module. The existing non-deployments had velocity changes of 0.40 and 0.44 mph. Analysis of their test data led these authors to conclude that "computation of the change in velocity using the crash pulse that results from a pedestrian impact crash does not correlate well with methods identified for car to car and car to barrier crash tests. The probable reason for this discrepancy is that a pedestrian impact is a short duration event and results in very low acceleration levels to the vehicle. This is not unexpected given that fact that there is a 10× to 20× weight differential between a pedestrian and a vehicle. Also, an SDM [Sensing and Diagnostic Module] on a non-deploy event will go 'back to sleep' after the acceleration falls below its threshold."

A study by Rose et al. [5.10] presented the data shown in **Figure 5.1**. This figure plots the error in the EDR-reported ΔV for staged collisions reported in 15 studies from the literature. The error in the EDR-reported ΔV was quantified in these studies by comparing it to the ΔV calculated from on-board, laboratory-grade accelerometers. The data in this figure are for full-overlap front or rear impacts where the vehicles experienced insignificant yaw rotation following the collision. The ΔV calculated from the laboratory accelerometers on the test vehicle is plotted on the horizontal axis, and the error in the EDR-reported ΔV is plotted on the vertical axis. A negative ΔV on the horizontal axis is a frontal impact, and a positive ΔV is a rear impact. The errors reported on the vertical axis are calculated from the ΔV magnitudes, such that a negative error is always an under-reporting of

the ΔV magnitude and a positive error is always an over-reporting of the ΔV magnitude.

The dashed black lines in **Figure 5.1** represent a window of ±10% error, the required accuracy per 49 CFR part 563.8 and the rule of thumb first found in the literature based on a study by Chidester's examination of early GM Sensing and Diagnostic Modules [5.11]. A significant number of the data points in this figure are from the NHTSA's New Car Assessment Program (NCAP) 35-mph frontal barrier crash tests, which yield ΔVs around 40 mph. The data show generally increasing absolute error in the EDR-reported ΔVs with increasing ΔV, consistent with the trend in the 10% error band. And most of the data do lie within the 10% error band. However, there are also points that fall outside of this window. In particular, for frontal collisions with ΔVs around 40 mph, there are points with considerably higher error than 10%. These high errors are often due to collision damage occurring in the area where the ACM is mounted. On the other end of the spectrum, for frontal collisions with ΔVs lower than 10 mph (more relevant to pedestrian collisions), almost all of the points lie outside of the 10% error window. For frontal impacts in this range, the EDR-reported ΔVs consistently underestimate the actual magnitude of ΔVs. Given the findings from the study by Fugger, Randles, and Eubanks combined with the results plotted in **Figure 5.1**, the best strategy for using EDR data for reconstructing a pedestrian collision would be to assume that the EDR-reported ΔV is not accurate and to focus on incorporating the EDR-reported pre-crash data into the reconstruction. This approach is not a problem for the reconstructionist since none of the widely used methods for pedestrian collision reconstruction would require knowing the striking vehicle's ΔV.

Figure 5.1 Error in the EDR-reported ΔV for full-overlap frontal impacts.

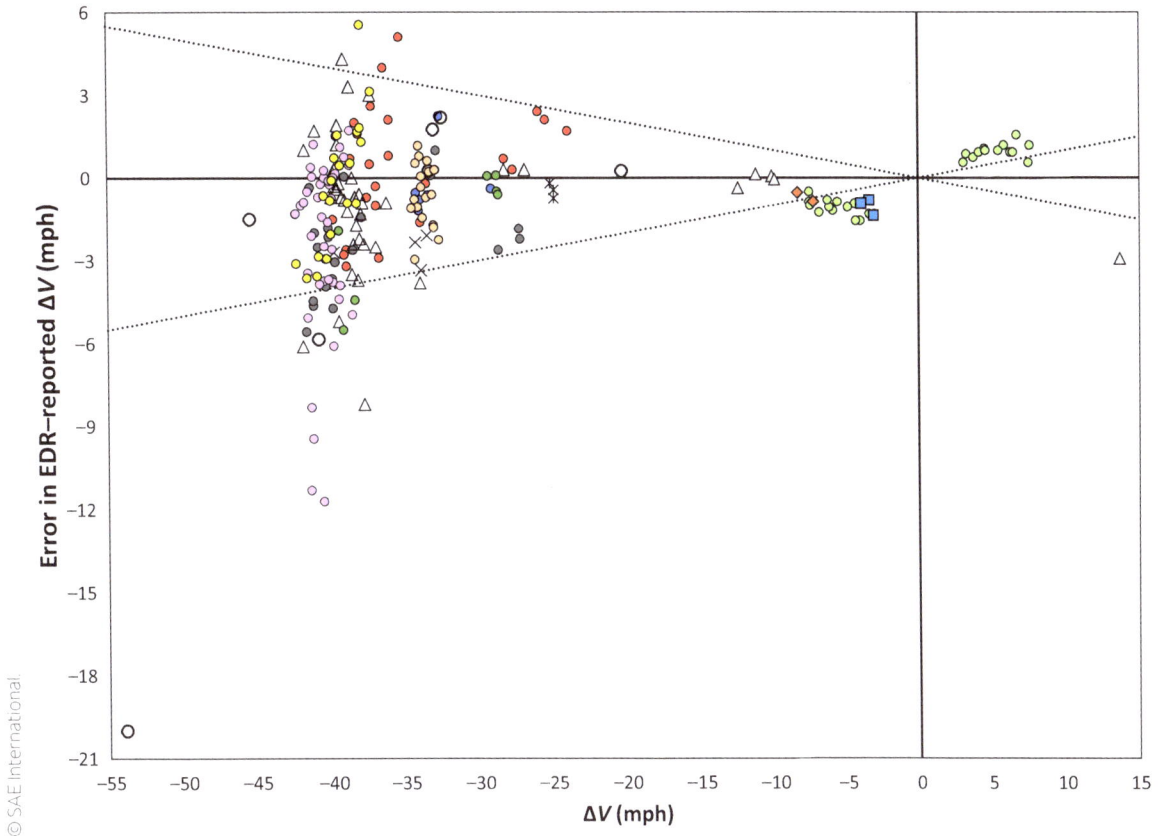

In 2016, Bortles et al. [5.12] presented a literature review related to the accuracy of the pre-crash speeds reported by passenger vehicle EDRs. They noted that the "EDR reported vehicle [pre-crash] speed is typically measured by sensors monitoring the output of the transmission or an average of the speed of the drive wheels. These sensors can accurately report wheel speed but, due to certain factors, the wheel speed may not represent the true over-the-ground speed of the vehicle. These factors may include longitudinal wheel slip due to acceleration or braking, wheel slip due to rotation of the vehicle about the vertical axis, significant changes in the tire's rolling radius as compared to the vehicle's original equipment, and changes to final drive ratio compared to the vehicle's original equipment." Each of these factors can be accounted for in the analysis, and the reconstructionist will typically be able to calculate the over-the-ground speed of the vehicle by making corrections for any of these factors that are relevant for a specific case. For most pedestrian collisions, the vehicle will not have developed significant sideslip at the time of the collision, and so rotation about the vertical axis of the vehicle will not be a factor that needs to be considered. Wheel slip due to braking and differences in tire size or final drive ratio are factors that could be relevant.

From their literature review, Bortles et al. concluded that, for straight, steady driving (no braking), the pre-crash speeds reported by the EDR will typically be lower than the actual speed but within 1 to 2 mph of the actual speed. Under heavy braking, the reported pre-crash speeds will typically be lower than the actual speed, but usually within 4 mph of the actual speed (or potentially more for vehicles not equipped with ABS). One challenge to using pre-crash speed data is determining precisely how the data sync to the crash. Suppose, for example, that an EDR reports 5 sec of pre-crash data at 0.5-sec intervals. How should one interpret the speed reported at time zero? This depends on the specific system, but for some systems, this reported speed at time zero will simply be the last speed reading collected by the ACM prior to the collision, which could be anytime between 0 and 0.499 sec prior to the impact. One way to handle this situation is to consider both ends of this range, considering the degree to which the vehicle is accelerating or decelerating over this 0.5-sec interval. Sometimes other evidence will inform a more precise syncing of the data to the collision. Some ACMs do sample pre-crash data elements at the time of the triggering event, and for these ACMs, this consideration would not be necessary. Another consideration with the pre-crash data is that sensors that measure and report pre-crash data elements can have full-scale values that are too limited to capture the full value of a data element. For example, some systems have a limit on the highest speed they can report, and others have a limit on the magnitude of the steering input they can report. These limitations are often listed in the data limitations section of the CDR report.

Toyota Vehicle Control History (VCH)

With the rise of pre-collision driver assistance systems, there is now the possibility that the activation of one of these systems—pedestrian AEB, perhaps—could result in a pedestrian collision being recognized and recorded by an EDR. Newer Toyota and Lexus vehicles, for example, will record non-collision events and report them as VCH records. These records include reporting of various parameters for 5 sec preceding the triggering event, with the reporting frequency depending on the type of event (i.e., hard braking, acceleration or steering, or activation of the AEB or ABS). Some event types include images captured by the forward-facing camera present on newer Toyotas.

A study by Xing, Yang, Tsuge et al. [5.13] assessed the accuracy of several parameters included in VCH records generated from AEB and ABS events. Their study utilized a 2017 Toyota Corolla with Safety Sense P Pre-Collision System (PCS). This system utilizes a radar sensor and a camera to detect and monitor the vehicle or pedestrians ahead [5.14]. The vehicle was driven toward a mock-up of a sedan at various speeds and accelerator pedal positions. The AEB system was allowed to activate. The speed, acceleration, closing speed, and target distance that were reported in the VCH records were compared to the values of these parameters documented using other instrumentation. These authors reported that the vehicle speeds in the VCH records that came from the vehicle speed sensor (VSS) consistently under-reported the reference speeds by -0.12 ± 0.16 m/s (-0.25 ± 0.36 mph). They noted that this under-reporting was, in part, due to the speeds being rounded to the next lowest km/h. In tests where the ABS did not activate, the under-reporting of the speeds was -0.08 ± 0.16 m/s. In tests where the ABS did activate, the under-reporting was -0.56 ± 0.69 m/s.

The distance to the vehicle ahead and the speed data reported from the radar consistently lagged behind the speed data from the other instrumentation. The lag increased with increasing acceleration or deceleration. The radar-measured distance to the vehicle ahead over-reported this distance measured with the other instrumentation by 1.18 ± 0.60 m. This over-reporting showed a trend of higher errors being associated with higher closing speeds. The radar-measured speed (closing speed) under-reported the actual speed while the vehicle was accelerating $(-0.36 \pm 0.16$ m/s) and over-reported the vehicle speed while decelerating $(1.32 \pm 0.84$ m/s).

To obtain sample VCH records, we imaged data from a 2021 Toyota RAV4 TRD Off-Road. The obtained data included several VCH records that contained images from the forward-facing camera of the vehicle. These images are referred to as freeze frame data (FFD). The reports for these events included a "General Information" section which noted that "VCH data is recorded in the memory of the airbag ECU…" Anecdotally, the authors have encountered an instance in which a Toyota vehicle was impacted, and the airbags deployed. The vehicle was then repaired [including replacing the airbag electronic control unit (ECU)] prior to the airbag ECU being

imaged. An imaging was then completed after the repairs. There were, of course, no data from the crash on the new airbag ECU. However, images from the crash were obtained, indicating that these images were not stored on the airbag ECU.

As an example of a VCH record, consider one of the VCH records that was obtained from the 2021 RAV4. This event was a sudden brake event. The VCH record for this event reported the odometer reading of the vehicle at the time of this event. Other parameters, including the vehicle speed, wheel speeds, the accelerator and throttle opening ratio (%), engine RPM, cruise control status, shift position, brake status, brake oil pressure, yaw rate, longitudinal and lateral acceleration, steering angle, vehicle stability control status, and ABS status, were reported at a rate of 6.7 Hz. The data were provided in both a tabular and graphical format. **Figure 5.2** is the vehicle speed graph from this event. Time zero is the hard brake trigger, and 5 sec of data are reported prior to the trigger and 5 sec after. The freeze frame images from this event were separated by 0.6 sec. **Figure 5.3** includes two sample freeze frame images. These images are of low resolution and black and white. However, roadway striping is discernible, as are the traffic signals and which signals are illuminated. It is not hard to imagine how such data and images could be invaluable when reconstructing a pedestrian crash.

Figure 5.2 VCH speed data from hard brake event.

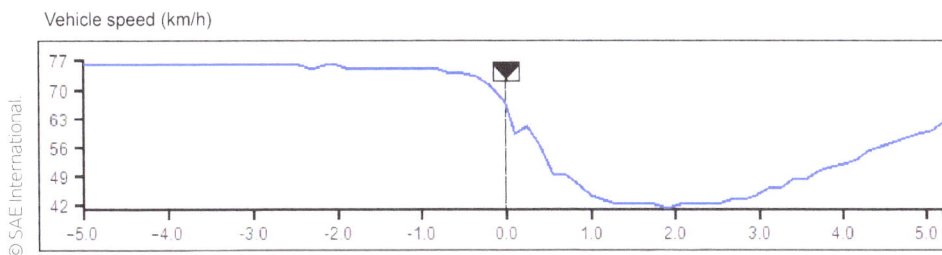

Figure 5.3 Freeze frame images.

(a)

(b)

© SAE International

Case Study: Related Event?

This crash occurred during the hours of darkness in a bustling and glamorous desert city. The collision involved a pedestrian and the left front of a Toyota Sienna minivan. Damage to the Toyota from this collision is depicted in **Figure 5.4**. As this photograph shows, damaged components at the left front of the Toyota included the bumper fascia, the headlight assembly, the hood, and the windshield. Airbags in the Toyota did not deploy. When this collision occurred, the roadway was dry, the weather was clear, the roadway was illuminated with "continuous roadway lighting," and the speed limit was 45 mph.

Figure 5.4 Collision damage to the Toyota Sienna.

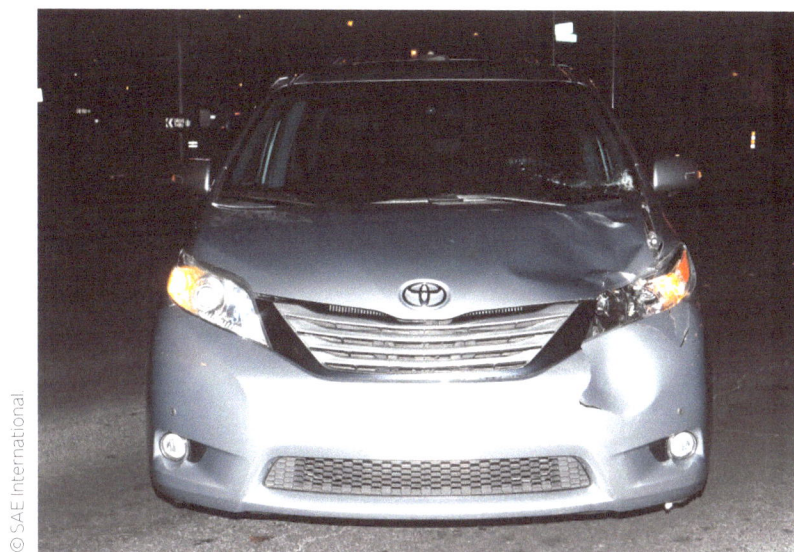

The Toyota was equipped with an EDR. The ACM on the Toyota, which contained the EDR, was a 12EDR, which according to the "General Information" in the CDR report was designed to be compatible with NHTSA's 49CFR Part 563 rule. During an inspection conducted after the vehicle had been repaired, data were retrieved from the ACM using the Bosch CDR system. A single event was recovered. Was this event from the subject crash? This question might arise often for pedestrian collisions. Given the significant weight discrepancy typically present between the vehicle and a struck pedestrian, EDR events from these collisions will often be non-deployments where the vehicle could have been repaired and driven a substantial distance after the crash. The recovered event was classified as a non-deployment side crash. The CDR report indicated that the satellite sensor in the front door sustained a maximum lateral ΔV of 0.3 mph, and the satellite sensor in the C-pillar sustained a maximum lateral of ΔV of 0.1 mph. The location and orientation of these velocity

changes was one clue that this event probably was not from the subject crash.

Beyond that, the system indicated that, at the time of the download, the vehicle had undergone 17,524 ignition cycles. At the time of the event, the vehicle had undergone 4258 ignition cycles—a difference of 13,266 ignition cycles. A study by Boots and Bartlett [5.15] concluded that vehicles would undergo an average of 4.97 ± 2.39 ignition cycles per day. Thus, in this case, the discrepancy between the ignition cycles at the time of the event and the time of the download would equate to a range of 1802 to 5142 days. The number of days between the subject collision and the date of the EDR download was 1082. Presumably, the Toyota would not have been on the road for all of these days, since it had to have been in the shop being repaired for at least some of those days. So, this range of days based on ignition cycles falls outside the range of days that would be indicated by the number of ignition cycles, supporting the interpretation that this event was not from the subject incident. Of course, as Boots and Bartlett noted, "just because there appears to be an ignition cycle 'discrepancy' does not mean you should ignore the data and/or immediately assume it is not from your crash." There can be vehicles with an outlier number of ignition cycles. Still, when this indicator is paired with other indicators, it supports the interpretation that the event is not from the subject crash. A vehicle history report was obtained for the subject vehicle. In addition to the subject crash, the vehicle history report indicated that the vehicle was involved in a rear-end collision without airbag deployment on a date 2918 days before the EDR data were downloaded. No further details were given in the history report, but this accident fits within the expected date based on the average daily ignition cycle range.

Finally, pre-crash data was reported at 0.5-sec intervals for approximately 5 sec prior to the non-deployment event. The vehicle indicated speed during this timeframe was a constant 37.3 mph (60 km/h) for the first six speed readings. The speed then increased to 38.5 mph (62 km/h) over the next two readings and then dropped to 36.7 mph (59 km/h) at the time of the event trigger. The pre-crash data indicated that the brake was never applied during the 5-sec recording window. The subject crash was captured by two cameras mounted on traffic signal poles near the area of the collision. The first video showed the Toyota completing a right turn onto the road where the collision occurred and then beginning to accelerate and drive toward the area of the collision. As it drove toward the area of the collision, the Toyota made a lane change from the right lane into the center lane. The second video showed the Toyota as it completed the right turn and traveled toward the camera. This footage also showed the pedestrian crossing the roadway, the Toyota contacting the

pedestrian, and the vehicle stopping. These videos were analyzed using methods detailed in Chapter 4. The first video showed the brake lights of the Toyota coming on just prior to the impact, which was inconsistent with the subject non-deployment event being from this crash. In addition to that, while the EDR-reported pre-crash speeds were of the same order of magnitude as the speeds obtained from the video, the trend of those speeds did not agree with the trend of the speeds obtained from the video. Ultimately, we concluded that the non-deployment event was not from the subject crash.

Case Study: Is this Event Related?

The previous case study can be juxtaposed with this one, where an EDR event turned out to be from the subject crash. This crash involved a later-model Hyundai sedan colliding with a pedestrian. In addition to the driver, the Hyundai was occupied by a passenger in the right front seat. It was dark, and there were no streetlights in the area. The roadway was dry, and the speed limit was 50 mph. The Hyundai exhibited significant damage to the hood and windshield, and there was also damage to the front license plate frame and the roof of the Hyundai. The pedestrian came to rest on the roadway, in front of the Hyundai. The driver of the Hyundai told investigating officers that he had his cruise control set at 50 mph. He stated that he did not see the pedestrian and braked when the collision occurred. The pedestrian was clothed in a black t-shirt and black shorts.

The Hyundai was equipped with ABS, electronic brake force distribution, and brake assist, but not with FCW or AEB. The vehicle had halogen bulb headlights. The vehicle was equipped with an EDR. The EDR was imaged, and one event was recovered. In this instance, the event was recorded at 11,816 ignition cycles and downloaded at 11,888 ignition cycles, a difference of 72 ignition cycles. According to the Data Limitation section in the Hyundai EDR report, the ignition cycle count will increment by one when the power mode is changed from OFF/Accessary to IGN ON/RUN and when the EDR is downloaded. During the inspection of this vehicle, the front bumper, hood, and roof had been repaired, but the windshield had not been. The owner of the repair shop communicated that the vehicle had been driven in and out of the garage during the repair process. This was a plausible explanation for the difference in ignition cycles.

The pre-crash data from the recorded event encompassed a timeframe of approximately 5 sec preceding the event, reported at 0.5-sec intervals. The vehicle-indicated speed during this timeframe was 78 or 79 km/h, which equates to approximately 49 mph. This is consistent with the involved driver's statement that he had his cruise control set at 50 mph, though the data do not specify whether or not the cruise control was active at the time of the collision. The data also indicated a clockwise (rightward) steering input of 5° at approximately the time of the event. This was consistent with the fact that, following the subject collision, the Hyundai came to rest angled slightly to the right. The pre-crash data indicated that there was no brake application in the 5 sec preceding the collision, though there is a 0.5-sec uncertainty in the data. The final point at 0.0 sec could have occurred as much as 0.5 sec prior to the collision.

A table from the EDR data called "System Status at Event" indicated that there was a driver and a passenger in the vehicle at the time of the event, and both were buckled. This is also consistent with the subject crash. Finally, the EDR data indicated that during the 300 ms of the recorded event, the vehicle experienced a longitudinal reduction in velocity of 10 km/h (6.2 mph). This is too high of a velocity change to have been caused by the impact with the pedestrian alone, and there was likely braking that contributed to this speed change. When combined with the distance the Hyundai traveled after the collision, this velocity change led us to conclude that the driver of the Hyundai applied emergency-level braking immediately prior to or at the time of the collision. With a deceleration of 0.76g, the vehicle would experience a velocity change just from braking during the 300-ms recording window of 4.9 mph. This level of braking would result in a braking distance of approximately 98 ft, which was consistent with the distance from impact to rest documented by the investigating officers. This confirmed that the EDR data could be attributed to the subject crash.

Case Study: Video Combined with Event Data

This case study considers a pedestrian collision that occurred during the hours of darkness in a construction zone in a state known for its orange juice. The weather was clear and dry at the time. The pedestrian was carrying a construction sign across an interstate off-ramp and was struck by a cement truck that was exiting the interstate. In this instance, the cement truck was equipped with a DriveCam system (now called Lytx), which is an aftermarket, in-vehicle, event-triggered video and data recorder that is typically mounted on the windshield a vehicle. These units contain a GPS sensor that measures speed, accelerometers that measure accelerations, and two cameras that record video. One of these cameras looks forward through the windshield, and the other looks rearward at the vehicle occupants. If the unit measures an acceleration that exceeds a preset threshold, an event is triggered and the unit stores video, acceleration, and speed data related to the event.

In this case, the raw data from the DriveCam was provided (file extension: dce). These data were viewed in the DriveCam Event Player. The frame of the data at the time of the triggering event is included in **Figure 5.5**. This frame is labeled with the time +0.00 sec. In this instance, the triggering event was heavy braking by the driver. The data extended from 8 sec prior to this trigger to 4 sec after the trigger. Times prior to the triggering event were labeled with negative times, and times after the triggering event were labeled with positive times. Video frames were captured at 4 Hz (0.25 sec). In the view of the DriveCam Event Player, a frame of forward-facing video was shown, as was a frame of interior-facing video. A speed of 50 mph was indicated, which was a speed reported from GPS. Instantaneous longitudinal and lateral accelerations were also indicated, and graphs of the accelerations and speeds were displayed at the top of the view. In addition to the DriveCam data, a sudden deceleration event from the truck's engine control module (ECM) was provided.

From the review of the video frames, the cement truck reached a stop at a time between +3.50 and +3.75 sec. However, at a time of +3.75 sec, the DriveCam event player continued to display a speed of 22 mph, even though the video showed the truck stopped. Thus, the updating and reporting of the GPS speed data lagged behind the video and acceleration data. This is consistent with what we have found in our research and experience related to the DriveCam and Lytx systems [5.16, 5.17]. We digitized the acceleration graphs from the DriveCam event player. This allowed us to access these accelerations with a

Figure 5.5 DriveCam Event Player view (time of triggering event).

greater frequency than the quarter-second intervals accessible on the display of video frames.

The longitudinal acceleration data were then adjusted and integrated to obtain speed in accordance with the methodology laid out in previous studies [5.16, 5.17]. The ECM data from the truck were also incorporated into this analysis. The graph in **Figure 5.6** plots the truck speeds as they were reported for the video frames in the DriveCam Event Player compared to the speeds obtained from integration of the adjusted longitudinal accelerations and from the sudden deceleration event, after they were properly synced to the video. As this graph shows, the GPS-reported speeds from the DriveCam were updating only once per second, and there was a

1- to 2-sec lag between the speeds when they were properly synced to the video and the speeds as they were reported in the DriveCam Event Player. That is the purpose of this case study—to highlight this syncing issue between the GPS speeds and the video. We have observed similar 1- to 2-sec lags on other cases, and in our research, and this is something that reconstructionists can be on the lookout for in evaluating speeds reported from GPS. The size of the lags we have observed has varied from case to case and between different versions of the DriveCam or Lytx Event Players. Such a lag is not present in every case, but similar lags have been observed in the speeds reported by the Geotab telematics system in commercial vehicle fleets [5.18] and also with GPS speed data from bicycle computers [5.19].

Figure 5.6 Comparison of DriveCam-reported speeds, speeds calculated from DriveCam accelerations, and speeds from the sudden deceleration event.

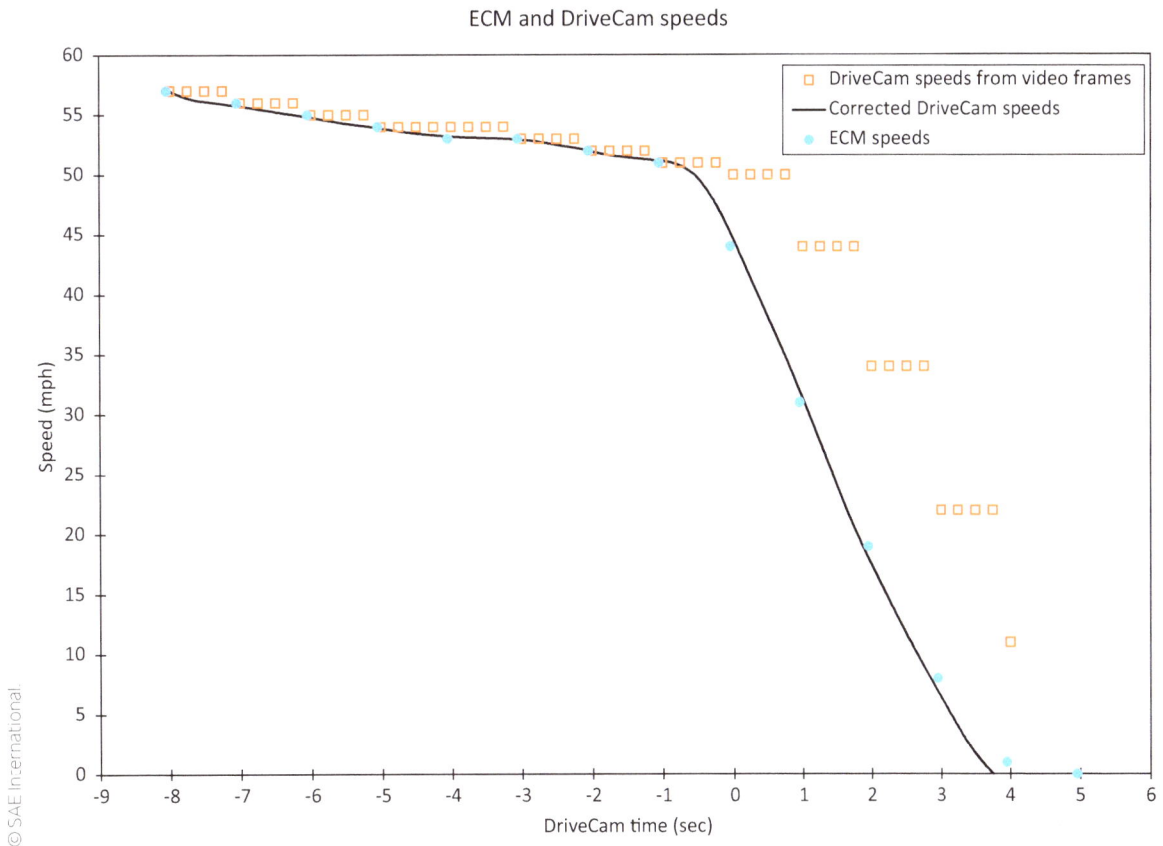

References

5.1.   Haight, R., "An Abbreviated History of CDR Technology," *Collision* 5, no. 1 (2010): 50-57, 64-73.

5.2.   Ruth, R., "Applying Automotive EDR Data to Traffic Crash Reconstruction," SAE Course #C1210, Course Slides.

5.3.   Wilkinson, C., Lawrence, J., Nelson, T., and Bowler, J., "The Accuracy and Sensitivity of 2005 to 2008 Toyota Corolla Event Data Recorders in Low-Speed Collisions," *SAE Int. J. Trans. Safety* 1, no. 2 (2013): 420-429, doi:https://doi.org/10.4271/2013-01-1268.

5.4.   Ruth, R. and Muir, B., "Longitudinal Delta V Offset between Front and Rear Crashes in 2007 Toyota Yaris Generation 04 EDR," SAE Technical Paper 2016-01-1496 (2016), doi:https://doi.org/10.4271/2016-01-1496.

5.5.   Xing, P., Lee, F., Flynn, T., Wilkinson, C. et al., "Comparison of the Accuracy and Sensitivity of Generation 1, 2 and 3 Toyota Event Data Recorders in Low-Speed Collisions," *SAE Int. J. Trans. Safety* 4, no. 1 (2016): 172-186, doi:https://doi.org/10.4271/2016-01-1494.

5.6.   Craig, V. (Eds), "N.H.T.S.A. Issues New Rule Regarding Event Data Recorders," *Accident Reconstruction Journal* 35, no. 1 (2025): 2-3.

5.7.   Chen, R.J., Tatem, W.M., Gabler, H.C., "Event Data Recorders (EDRs) Duration Study," Submitted to NHTSA, September 2017.

5.8.   Watson, R., Bonugli, E., Greenston, M., Santos, E. et al., "Event Data Recorder Trigger Probability in the Crash Investigation Sampling System Database," SAE Technical Paper 2024-01-5027 (2024), doi:https://doi.org/10.4271/2024-01-5027.

5.9.   Fugger, T., Randles, B., and Eubanks, J., "The Efficacy of Event Data Recorders in Pedestrian-Related Accidents," SAE Technical Paper 2004-01-1195 (2004), doi:https://doi.org/10.4271/2004-01-1195.

5.10. Rose, N. et al., "Motorcycle Accident Reconstruction: Applicable Error Rates for Struck Vehicle EDR-Reported ΔV," *Collision: The International Compendium of Crash Research* 13, no. 1 (2019): 88-111.

5.11. Chidester, A., Hinch, J., Mercer, T., and Schultz, K., "Recording Automotive Crash Event Data," in *Proceedings of the International Symposium on Transportation Recorders*, Arlington, VA, 1999.

5.12. Bortles, W., Biever, W., Carter, N., and Smith, C., "A Compendium of Passenger Vehicle Event Data Recorder Literature and Analysis of Validation Studies," SAE Technical Paper 2016-01-1497 (2016), doi:https://doi.org/10.4271/2016-01-1497.

5.13. Xing, P., Yang, M., Tsuge, B., Flynn, T. et al., "The Accuracy of Toyota Vehicle Control History Data during Autonomous Emergency Braking," SAE Technical Paper 2018-01-1441 (2018), doi:https://doi.org/10.4271/2018-01-1441.

5.14. Lewis, L., Hare, B., Clyde, H., and Landis, R., "Vehicle Control History: Data from Driver Input and Pre-Collision System Activation Events on Toyota Vehicles," SAE Technical Paper 2019-01-5094 (2019), doi:https://doi.org/10.4271/2019-01-5094.

5.15. Boots, K.E. and Bartlett, W., "Average Daily Ignition Cycles – An Update," *Accident Reconstruction Journal* (2023): 22-26.

5.16. Rose, N., Carter, N., Pentecost, D., Bortles, W. et al., "Using Data from a DriveCam Event Record to Reconstruct a Vehicle-to-Vehicle Impact," SAE Technical Paper 2013-01-0778 (2013), doi:https://doi.org/10.4271/2013-01-0778.

5.17. Rose, N.A. et al., "Using Data from a DriveCam Video Event Recorder to Reconstruct a Hard Braking Event," *Collision: The International Compendium for Crash Research* 7, no. 1 (2012): 110-120.

5.18. Steiner, J., Armstrong, C., Kress, T., Walli, T. et al., "Commercial Vehicle Global Positioning System Based Telematics Data Characteristics and Limitations," SAE Technical Paper 2017-01-1439 (2017), doi:https://doi.org/10.4271/2017-01-1439.

5.19. Siddiqui, O., DiBiase, S., Hoang, R., Nguyen, B. et al., "Evaluating the Accuracy and Reliability of Bicycle GPS Devices," SAE Technical Paper 2021-01-0882 (2021), doi:https://doi.org/10.4271/2021-01-0882.

Chapter 06

Human Factors and Nighttime Visibility

At this juncture, a review of the basic questions often asked when reconstructing a pedestrian collision will be useful.

1. Where on the road did the collision occur?

2. What was the impact speed of the vehicle?

3. How fast was the pedestrian moving at the time of the collision? Were they walking or running?

4. Are the descriptions of the crash given by witnesses and involved parties accurate?

5. Were there visibility obstructions that contributed to the crash?

6. Was the pedestrian visible in time for the driver to have responded to avoid the collision?

7. What actions would have been needed by the driver to avoid the pedestrian?

8. What could the pedestrian have done to avoid the collision?

9. At what time did the collision occur?

10. What were the lighting conditions at the time of the collision?

Previous chapters have detailed methods for answering some of these questions, but questions remain related to the visibility of pedestrians and the factors to be evaluated in assessing a driver's ability to avoid a collision with a pedestrian. Often, pedestrian collisions will involve some level of limited visibility, either from geometric obstructions (a parked vehicle, for instance) or from low light or nighttime conditions. This chapter will address these situations.

During daylight conditions, evaluating a pedestrian's visibility to a driver, or conversely, an approaching vehicle's visibility to a pedestrian, is often a matter of pure geometric visibility. The analyst determines whether physical objects were obstructing the view of either party and when those objects would have ceased to obstruct a view relative to when the collision occurred. This analysis is often independent of the distance separating one of the parties from the other, except for specific cases such as fog, dust, or smoke. Under night or low-light conditions, however, the light that reaches a pedestrian and is reflected from the pedestrian to a driver may not be sufficient to make the pedestrian detectable to a driver. The distance from the driver to the pedestrian through time is relevant, and the lighting conditions must be considered in addition to geometry obstructions.

The visibility of a pedestrian at night for an approaching driver will be affected by the illumination from the headlamps of the vehicle, by streetlamps, by ambient lighting, by oncoming traffic, and by characteristics of the weather. Primarily, it is the luminance contrast between an object and its background that makes an object visible to an observer at night [6.1]. Luminance is the amount of light that is reflected off a surface, and luminance contrast, or difference in luminance between two objects, is dependent on surface properties such as color and reflectivity, as well as the amount of illumination arriving at that surface by light sources. In reconstructing a nighttime crash, the analyst may need to evaluate the limits of visibility to determine the distance from which a pedestrian is visible. A number of authors have published methods for evaluating the limits of visibility by replicating the conditions present at the time of the crash and performing an *in situ* evaluation, through observation and light measurement [6.2, 6.3, 6.4].

Measuring Light

Photometry is the branch of science dedicated to measuring the intensity and characteristics of light, specifically on the human eye. A full treatment of photometry is beyond the scope of this book, but the treatment presented by Green et al. [6.5] is useful for a general understanding. There are several concepts in photometry that are worthy of discussion in the context of pedestrian accident reconstruction. These concepts are often utilized in modeling the visibility or detectability of pedestrians.

Illuminance

Illuminance (E) is the measure of the quantity of visible light that falls onto a given surface area. Illuminance at a surface depends on both the intensity of the light source and the distance of that surface from the light source. The SI unit for illuminance is lux (lx), which is equivalent to lumens per square meter, and the non-SI unit is the foot-candle. For reference, illuminance at ground level on a sunny day is approximately 100,000 lx, while nighttime illuminance in an area away from light sources and city lights is often below 1 lx. Illuminance is measured with an illuminance meter, sometimes called a lux meter. A commonly used illuminance meter is the Konica-Minolta T-10A (**Figure 6.2**).

For pedestrian accident reconstruction, illuminance meters are often used to measure natural ambient lighting at a scene, ambient lighting due to artificial light sources, and lighting due to headlights. Illuminance from separate light sources is additive, such that if multiple light sources are illuminating a surface, the total illuminance can be calculated by adding the illuminance contributed by each individual source. The inverse square law governs the relationship between illuminance and distance from a light source. Thus, if the illuminance from a light source is measured at a certain distance from the light source, the inverse square law can be used to calculate the illuminance from that source at a different distance. This is expressed in Equation (6.1). In this equation, E is the illuminance, measured in lux or foot-candles, I represents the light intensity at the source, measured in candelas, and d represents the distance from the light source.

$$E = \frac{I}{d^2} \tag{6.1}$$

Luminance

Luminance (L) is a measure of the quantity of light reflected from a surface area and can be thought of as how bright that surface appears to an observer. Luminance of a surface is dependent on both the amount of light falling onto that surface (illuminance) and the reflectivity of that surface. Luminance can be measured using a spot luminance meter, such as the Konica Minolta CS-150, or using an imaging photometer, such as the Westboro Photonics P501U. Luminance is measured in units of candelas per square meter (cd/m²). In assessing the visibility of a pedestrian, luminance measurements can be taken from inside a vehicle to quantify the light reflected from that pedestrian to an observer (often a driver) and the light reflected by the background surrounding the pedestrian. These measurements form the basis for luminance contrast, which is a useful concept for assessing visibility and is used in the Adrian visibility model, which is discussed below.

Reflectance

A surface that is illuminated through an external light source will reflect a portion of that light. Reflectance (ρ) is the ratio of reflected light to incident light. The reflectance of surfaces will vary, with flat black-painted surfaces reflecting very little light and retroreflective surfaces reflecting nearly all the incident light. Reflectance can be measured and quantified as expressed in Equation (6.2). In this equation, ρ is the reflectance, expressed as a ratio or percentage, L is the measured luminance of a surface, in candelas per square meter, and E is the measured illuminance, in lux.

$$\rho = \frac{\pi * L}{E} \tag{6.2}$$

To illustrate the concepts of illuminance, luminance, and reflectance, consider the photographs in **Figures 6.1** through **6.3**. In the depicted experiment, a single sealed beam motorcycle headlamp was energized by a 12-V direct current source, and the light was placed perpendicular to a wall in a dark lab. A standard 18% middle gray balance card was attached to the wall. This target is a standardized target used in photography to calibrate exposure levels. A Konica-Minolta T-10A was used to measure the illuminance at the target. The illuminance from the headlight was measured at 1727 lx, as shown in **Figure 6.2**. Next, a Westboro Photonics P501U imaging photometer was placed near the headlight, and the luminance of the card was measured with Photometrica software. An area of interest was created encompassing the gray balance card. This area of interest consisted of 62,620 measurement pixels at an average luminance of 100.9 cd/m². Entering the measured luminance and illuminance values into Equation (6.2) results in a calculated reflectance of 18%. The concepts of luminance, illuminance, and reflectance will be used in several models that can be used to evaluate when a pedestrian becomes detectable to a driver.

Figure 6.1 Experiment setup, single motorcycle headlamp.

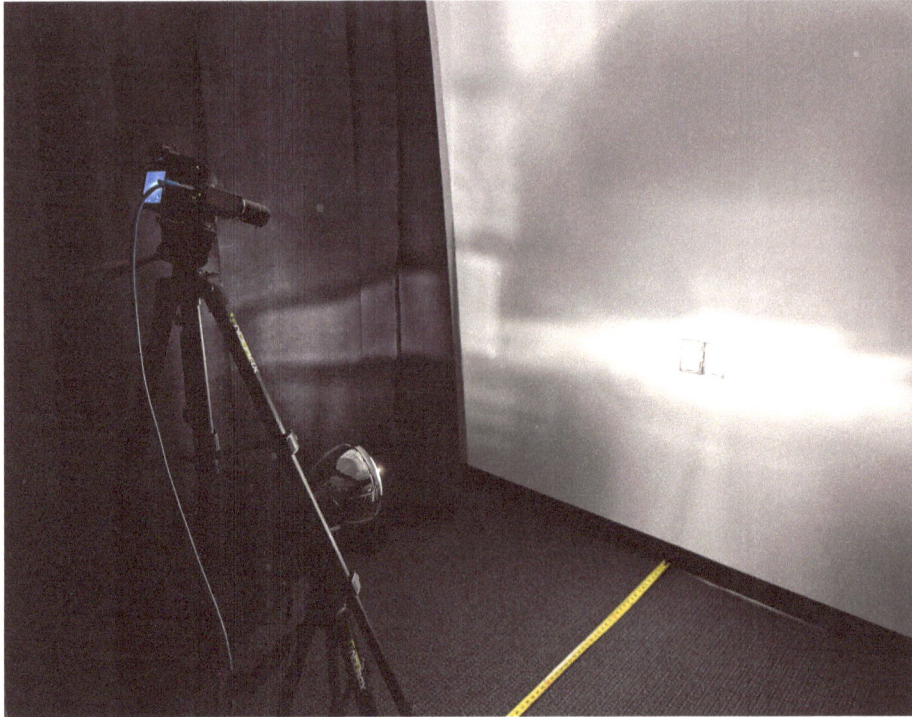

Figure 6.2 Illuminance measurement of 18% gray balance card.

Figure 6.3 False color luminance measurement with area of interest highlighted.

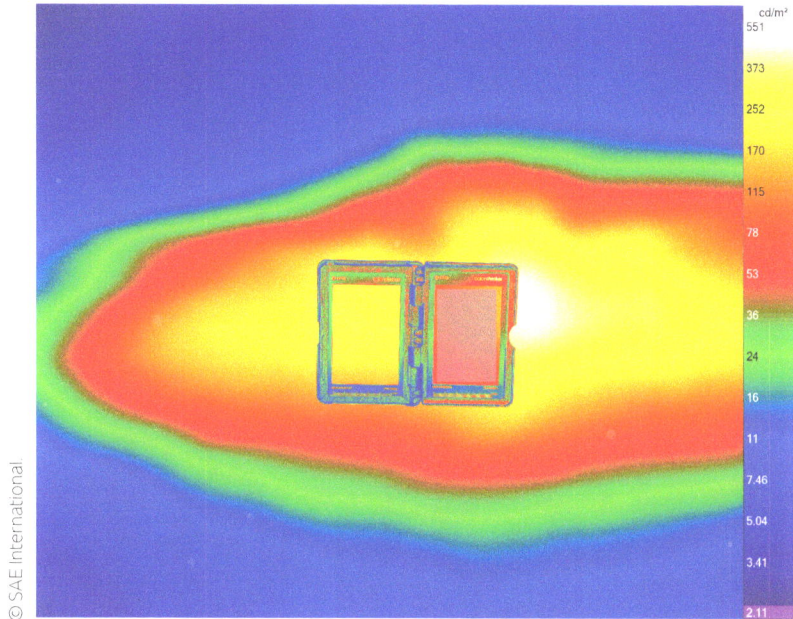

Sources of Light

Ambient Light

During daytime conditions on a roadway, the majority of illumination comes from natural lighting via the sun. As the sun sets, the lighting conditions change from daytime to twilight. Twilight refers to the period after sunset and before sunrise during which the atmosphere is partially illuminated by the sun. Early astronomers defined civil twilight as the time period after sunset or before sunrise when normal outdoor activities could be conducted without artificial illumination. Civil twilight is currently defined as the period of time when the geometric center of the sun is from 0 to 6° below the horizon. Civil twilight occurs in the evening after sunset (civil dusk) as the sun descends from the horizon to 6° below the horizon and in the morning before sunrise (civil dawn) as the sun ascends from 6° below the horizon to the horizon. Similarly, nautical twilight refers to the period when the geometric center of the sun is between 6 and 12° below the horizon, and astronomical twilight refers to the period when the geometric center of the sun is between 12 and 18° below the horizon. The diagram in **Figure 6.4** depicts these phases.

Figure 6.4 Twilight phases.

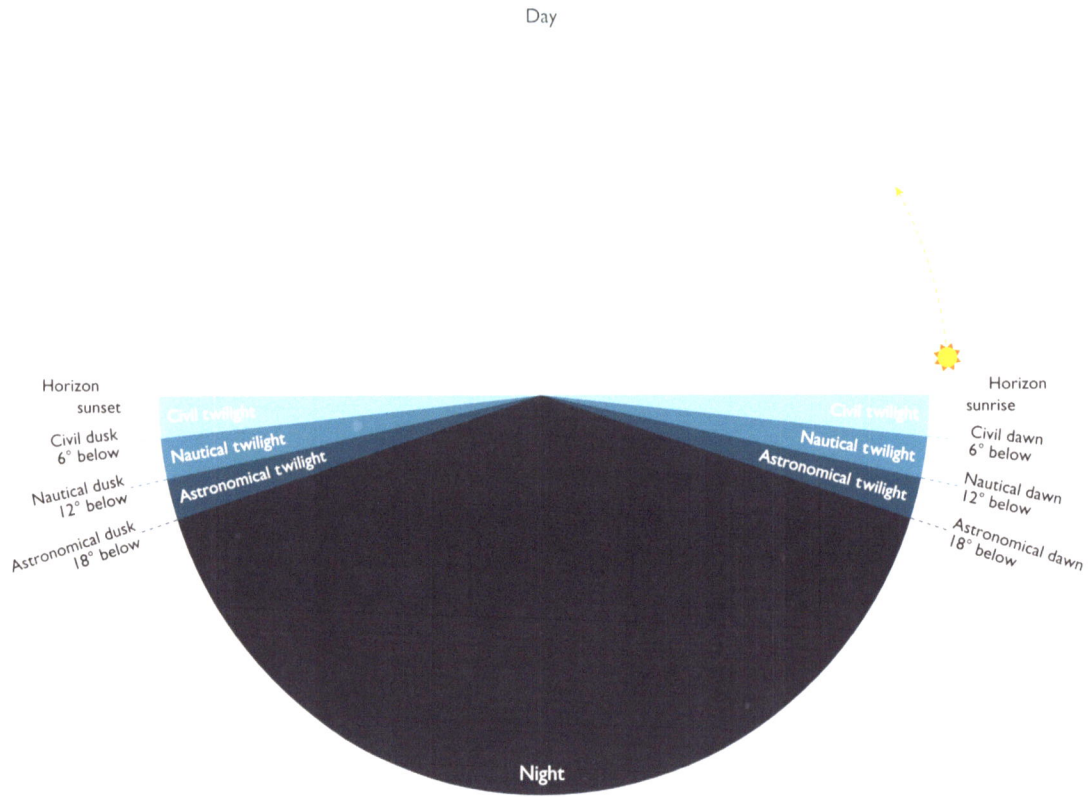

The reconstructionist may be interested in evaluating the ambient lighting that was present when a crash occurred. This necessarily involves an accurate assessment of the time at which the crash occurred. Police records, fire records, dispatch records, and timestamped videos may all give an indication as to the time of the crash. The sun's illumination changes most rapidly during civil twilight and civil dawn, so the sun's illumination during these phases will be particularly sensitive to the determined time of the accident.

Often, a site inspection or study at a time with similar ambient lighting conditions is desired. The use of a sun calculator will inform the analyst of the altitude and azimuth of the sun at a given location and at a given time. The Astronomical Applications Department of the United States Naval Observatory provides one such calculator (https://aa.usno.navy.mil/data/AltAz), as does the website suncalc.org. Adequate matches of the solar altitude and azimuth can be achieved around the anniversary date of the crash, as well as on days equidistant from the summer solstice (June 20 or 21) and winter solstice (December 20 or 21).

Auxiliary Lighting Sources

In addition to lighting from the sun, a crash scene may be illuminated by nearby light sources, such as overhead lights, illuminated signs, and lights from nearby buildings. These auxiliary light sources can be documented and accounted for in an accident reconstruction. Often, photographs taken near the time of the accident will document these additional lighting sources. When evaluating external lighting sources, take note of which lighting sources were activated at the time of the crash. Lamps will sometimes burn out between the time of the accident and the time of a site inspection, or they will have been burnt out at the time of the crash and replaced by the time of the inspection. Take note of the type of light as well. Prior to the prominence of LED lighting, many overhead streetlights were sodium vapor lights. These lights are gas discharge lamps that generate light by passing a current through vaporized sodium, rather than by heating a metal filament. Many municipalities are transitioning from sodium vapor streetlights to LED lighting, given the energy efficiency improvements of LED lighting systems. A situation may arise where the accident occurred under sodium vapor street-lights and at the time of the inspection those lights have been upgraded to LED lighting. The two lighting types are usually distinguishable by the color temperature of the lights. Sodium vapor lighting has a warm yellow color (about 2000K), whereas LED lighting has a whiter or bluer appearance (around 3000 to 4000K).

Vehicle Headlamps

The primary source of illumination in a nighttime pedestrian crash is sometimes the headlamps of a driven vehicle. Thus, an under-standing of vehicle headlamp illumination patterns is often useful in a nighttime pedestrian accident reconstruction. As a vehicle approaches, the headlamps will illuminate the pedestrian to provide contrast to the driver between the pedestrian and the background. A driver's ability to avoid a pedestrian at night will often be dependent on the illumination characteristics of the vehicle headlamps, the shade of the

pedestrian's clothing, and the speed of the approaching vehicle. Headlamps range from light types such as incandescent, HID, to LED and include sophisticated reflectors and lenses that shape and distribute light in a wide variety of ways. Each headlamp model has a unique beam pattern. In a 2013 paper, Muttart et al. [6.6] mapped the headlamp illuminance of various vehicles and noted the headlamp type and age of the vehicle. The authors then developed empirical relationships to provide illumination estimates at longitudinal and lateral distances from the vehicle for low beams and high beams as a function of headlamp type (motorcycle, halogen reflector, halogen projector, LED reflector, LED projector, HID reflector, HID projector), vehicle type (motorcycle, passenger vehicle, large vehicle), and headlight age (dirty and/or greater than seven years old, 50th percentile, clean and less than three years old). These empirical equations are incorporated into Response software and provide an estimate for a vehicle headlamp illumination distribution [6.7].

Depending on the specifics of the crash, a reconstructionist may want to map the headlamp beam pattern from the accident vehicle or an exemplar, rather than utilizing these empirical relationships. Vehicle-specific headlamps can be mapped by measuring headlamp illuminance in a grid pattern at various distances in front of and lateral to a vehicle. Headlamps can also be mapped by measuring distances in front of and lateral to a vehicle where threshold levels of illumination (say, 5, 10, 15, and 20 lx) are met. Both methods require a dark, flat, wide-open space. In a 2021 paper, Funk et al. [6.8] proposed the use of a robotic total station and an illumi-nance meter to optimize the process of mapping headlamp illumination patterns. This method creates a datum based on the vehicle plane to ensure that the illumination measurements are taken at a consistent height relative to the vehicle. The proposed method provides a consistent headlamp mapping procedure and minimizes mapping time while only requiring one person. In a follow-up paper, Funk et al. [6.9] compared empirical predictions of headlamp illumination levels to actual headlamp illumination maps. The authors found that, while actual illumination distances fell within the range of the empirical prediction, mapping the headlamp pattern of a specific vehicle provides additional precision that can be useful in a reconstruction. **Figure 6.5** is from the Funk et al. study and depicts the illumination pattern at various levels for a 2011 Toyota Sienna with HID bulbs. In this figure, the longitudinal and lateral distances from the front left corner of the vehicle are depicted on the x-axis and y-axis, respectively, in meters.

Figure 6.5 Headlamp illumination distribution graphic [6.8].

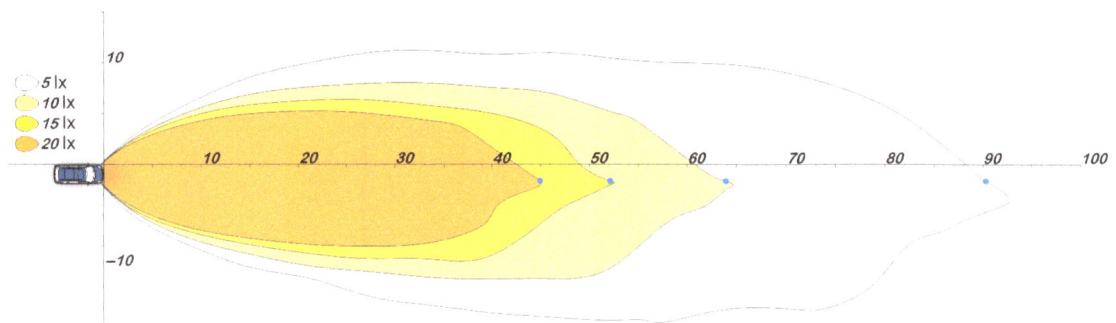

Reprinted from Reference [6.8]. © SAE International.

Assessing the Nighttime Visibility of Pedestrians

When a reconstructionist is evaluating a driver's actions and performance in the moments leading up to a collision with a pedestrian, the following question arises: When would an attentive driver have been able to detect the presence of the pedestrian? The likelihood that a driver will detect a pedestrian is influenced by the contrast between the pedestrian and their surroundings, the driver's light adaptation level, the size of the pedestrian, the driver's level of anticipation and expectancy, and the time available for viewing the pedestrian. Contrast is defined as the amount of light (luminance) reflecting from a subject compared to the luminance of the area surrounding that subject. In some situations, headlights of a vehicle or other lighting sources will illuminate a subject so that it is brighter than its background, which is known as positive contrast. An example of positive contrast is depicted in **Figure 6.6**. Other times, the background of the subject will appear brighter than the subject itself, which is known as negative contrast. An example of negative contrast is depicted in **Figure 6.7**. Negative contrast sometimes occurs when a pedestrian crosses in front of the headlights of a vehicle.

Figure 6.6 Pedestrians crossing a roadway under positive contrast conditions.

Figure 6.7 Pedestrians crossing a roadway under negative contrast conditions.

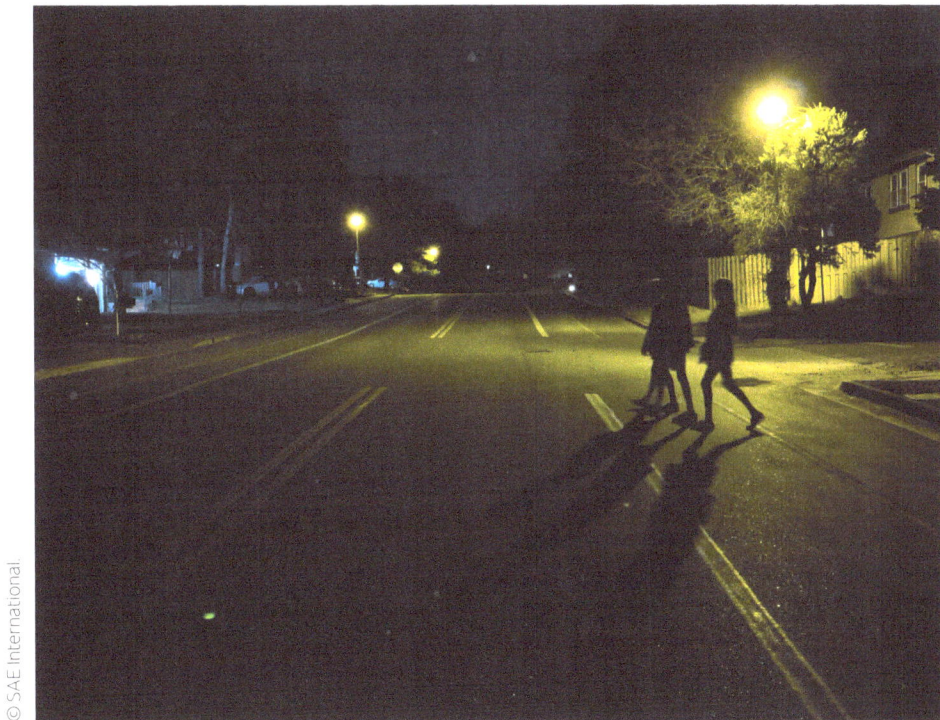

Researchers have presented empirical and theoretical models to evaluate object visibility under nighttime or low-light conditions. In a 1989 paper, Owens, Francis, and Leibowitz [6.4] proposed a model that uses the concept of civil twilight as a threshold for the visibility of an object. Generally, during civil twilight, many objects are discernible without artificial lighting. The concept here is that if artificial lighting illuminated an object to the same extent as the object was illuminated at the end of civil dusk, the object would be visible. Owens et al. noted that natural ambient illumination is 30 foot-candles (323 lx) at the beginning of civil dusk and 0.3 foot-candles (3.2 lx) at the end of civil dusk. Thus, Owens et al. used a threshold value of 0.3 foot-candles and proposed a "twilight distance" (TD) in which a headlight or headlights provide illumination of 0.3 foot-candles to an object.

This distance is represented by Equation (6.3), in which *TD* is the twilight distance in feet, *CD* is the luminous intensity of the light source in candelas, and the coefficient 0.3 represents the illumination at the end of civil dusk in foot-candles. This simplified model does not account for the reflectivity of the object of interest or the contrast of the object to the background:

$$TD = \sqrt{\frac{CD}{0.3}} \tag{6.3}$$

In practice, it is unlikely that one would measure the luminous intensity of a vehicle headlight system to calculate the TD. However, the same concept can be applied by measuring the illuminance of headlights on the subject or exemplar vehicle and determining where the 3.2-lx threshold occurs in relation to the vehicle. Subsequent work by other researchers has proposed illumination thresholds higher than 3.2 lx for a driver to detect a pedestrian.

Several other models have been proposed to evaluate the detectability of pedestrians. Perhaps the most comprehensive model is often referred to as the Adrian visibility model, which accounts for target and background luminance, target size, motion, visual field location, viewing duration, glare level, observer adaptation, expectancy, fatigue, and age. The inputs to this model are not always known in a reconstruction setting, but the model still provides a useful framework for understanding human detection of objects in nighttime conditions. Another model, which will be referred to as the "illumination threshold model," evaluates the time at which a pedestrian is subject to a specific level of illuminance, typically from vehicle headlights. The threshold level used is based on the shade of clothing that the pedestrian is wearing.

Adrian Visibility Model

Adrian presented his model in a 1989 paper [6.2]. This model quantifies visibility based on the contrast between an object and its background, the size of the object, the reflectivity of the object, contrast polarity (positive or negative contrast), observer age, and exposure time. This model was based on previous work by Adrian [6.10, 6.11] and others, including Aulhorn [6.12] and Blackwell [6.13, 6.14]. The model allows for the calculation of luminance difference (ΔL) thresholds. Luminance difference is simply the difference between the luminance of a target object and the background of that object, as shown in Equation (6.4). The thresholds proposed by Adrian represent the luminance difference between an object and its background at which nearly all observers (99.93%) would detect the object <u>under laboratory conditions</u>.

$$\Delta L = L_o - L_b \tag{6.4}$$

Adrian further proposed that under real-world conditions, a multiple of ΔL would be necessary depending on the visual task demand. Adrian used the term *visibility level* (VL) to define the multiple of the calculated ΔL required to render conspicuity to a target. VL is defined as the ratio of actual luminance difference to threshold luminance difference, as shown in Equation (6.5). In the luminance range of street lighting, Adrian proposed a VL range between 10 and 20. Adrian found that higher luminance difference levels were generally required to generate positive contrast than what were required to generate negative contrast:

$$VL = \frac{\Delta L_{actual}}{\Delta L_{threshold}} \tag{6.5}$$

Quantitatively, the contrast between an object and its background is defined as the difference between the background luminance and the target luminance divided by the background luminance, as shown in Equation (6.6):

$$C = \frac{L_o - L_b}{L_b} = \frac{\Delta L}{L_b} \tag{6.6}$$

In 2003, Ising [6.15] applied the Adrian visibility model to drivers to determine threshold VLs for perceiving an object on the road ahead while driving at night. Ising noted that Adrian's model is a useful tool to assess the visibility of an object at night, but it was developed under laboratory conditions and drivers operating motor vehicles on the road require VLs greater than those needed in the laboratory. Ising noted that, while Adrian proposed a VL between 10 and 20, the proposal was based on data from only young laboratory observers and the visual acuity necessary to read roadway signs. Ising analyzed experimental data on driver detection distances of pedestrians from the Olson experiment [6.1] and compared them to Adrian's model. In the Olson testing, drivers were asked to verbally announce when they detected a pedestrian target as they drove toward the targets at a speed of 25 mph. Ising accounted for the delay between the driver's detection and vocalization and the delay between the driver's vocalization and the experimenter's button-push response. Based on the detection distance, headlight illumination, target reflectivity, target size, and driver age, Ising calculated VLs for pedestrian-sized targets at the side of the road in the range of 1 to 23. However, the test subjects in this Olson experiment knew that they were being tested and were told on which side of the road a target would appear. Ising used the results of a study by Roper and Howard [6.16] to adjust the response distance to apply to unalerted drivers. Once this adjustment for expectancy was made, Ising reported VLs for unalerted drivers between 13 and 210. Olson noted that the data used to correct the alerted drivers to unalerted drivers was based on limited data with a single 1930 vintage headlamp, so the proposed VLs for unalerted drivers should be interpreted cautiously.

In a 2008 follow-up study, Ising [6.17] incorporated the CIE General Disability Glare Equation (CIE 2002) into the Adrian visibility model, resulting in a model that accounts for a wider range of glare source positioning. Ising again analyzed experimental data by Olson et al. [6.1] and evaluated VLs using the updated model. Ising reported that age, headlight beam pattern, and target reflectivity had a significant effect on VL at target detection. Ising found average threshold VLs between 14 and 89 for unalerted drivers and noted that "these ranges also illustrate a general weakness in the visibility level approach in that the detected targets did not have equal VLs in all cases. That is, equal VLs did not correspond to equal levels of task performance."

Illuminance Threshold Model

Some pedestrian collisions occur on dark, unlit roadways with minimal background luminance. In this scenario, where the only appreciable light source is the vehicle headlights, the only variable changing in Equation (6.6) is the luminance of the pedestrian as it is being illuminated by the vehicle headlight system. Thus, the only variables affecting contrast for a given vehicle–pedestrian pair are the pedestrian position relative to the headlight system and the reflectance of the pedestrian and clothing. In a 2013 publication, Muttart et al. [6.6] presented a model that correlates headlight illumination with driver response distances. The model requires two elements—an estimate of headlight luminance as a function of the pedestrian distance ahead of and lateral to the vehicle headlight system, and an estimate of when drivers will respond to pedestrians that are being illuminated by the headlight system based on the reflectance of the pedestrian and their clothing. These elements will be addressed separately below.

To determine the distance at which a given illumination threshold is met, a model of headlight illumination as a function of longitudinal and lateral distance from the vehicle is necessary. Muttart et al. mapped a total of 48 low-beam headlight systems, 26 high-beam headlight systems, seven vehicles with only the driver's side headlight illuminated, seven vehicles with only the passenger's side headlight illuminated, and five motorcycle headlights. Illuminance measurements were taken at heights of 0, 0.6 m (24 in.), and 1 m (39 in.). The vehicle age and headlight type were noted. Muttart then developed empirical models to estimate 3.2 lx (0.3 foot-candle) illumination distances as a function of headlight type, vehicle age, and lateral distance from the vehicle. The inverse square law—Equation (6.1)—along with an atmospheric reduction factor was then used to calculate the distance at which other illumination levels were met.

To correlate headlight illumination levels with driver response distances, Muttart et al. analyzed the results of 25 nighttime driver response distance experiments. They analyzed factors that correlate with a change in driver response distance, such as the lateral position of the pedestrian from the centerline of the vehicle, the headlight type, the positioning of the pedestrian (standing or slumped/laying on the roadway), and the shade of clothing that the pedestrian is wearing. They determined a threshold value for headlight illuminance based on the shade of clothing that the pedestrian was wearing and the relative positioning of the pedestrian and the vehicle, as well as the performance characteristics of the headlights and when the headlights would provide the threshold level of illuminance.

In the examined studies, drivers or passengers were asked to press a button, make a verbal response, or brake when they identified a pedestrian or other object. Muttart et al. analyzed the independent variables in these studies, including the experiment type (closed course or road or traffic study), clothing shade, movement of the pedestrian, lateral position of the subject relative to the vehicle, whether the subject responded when the pedestrian was first visible or when the pedestrian was recognized as a pedestrian,

whether the low beams or high beams of the vehicle were illuminated, vehicle headlamp type, whether the subject was driving or was a passenger, size of the pedestrian or object, and whether the subject had been familiarized with the target. These variables were coded to address the factors that might influence the distance at which a driver responds to an object, namely contrast, lighting, adaptation, pattern, and size. Based on the results of these studies, Muttart et al. ran a multiple linear backward regression to correlate the independent variables studied with the driver response distance. They noted that the shade of clothing had a significant influence on the driver's detection distance. Drivers responded earlier to lighter-colored (higher reflectance) objects. Objects to the left were recognized later than objects on the right, as were odd shapes or any targets when the vehicle was on a curve. Newer headlights were moderately associated with improved recognition. Based on the results of these studies, empirical curves were developed to estimate the illumination necessary to recognize pedestrians wearing clothing of various shades. Muttart et al. noted that previous research has shown that drivers need an illumination of 15 to 20 lx to recognize a darkly clad object, approximately 3.2 to 5 lx to respond to a grayish object, and 1 to 2 lx to recognize a lighter object. They summarized the illumination values at driver response as a function of the pedestrian clothing shade and prepared a chart, which is presented in **Figure 6.8**.

Figure 6.8 Illumination levels at driver response for differing pedestrian clothing shades, reproduced from Muttart (2013).

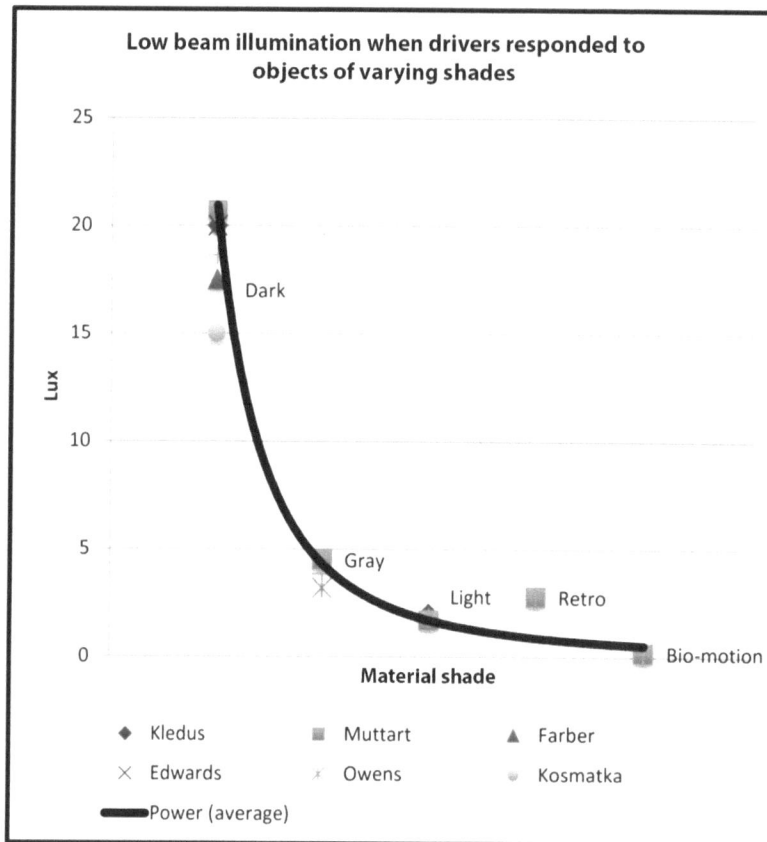

Combining the two above models—a headlight model that is dependent on vehicle characteristics and a recognition model that accounts for pedestrian clothing shade and driver responses—allows for a reconstructionist to estimate when a driver would be expected to react to a pedestrian in a path intrusion scenario. These models are incorporated into a software package called Response. The fidelity of this model may be improved when using vehicle-specific headlight mapping data, as opposed to the generic headlight illuminance model. An analyst can map the headlights of the subject vehicle or a suitable exemplar vehicle [6.8] and

incorporate the results into the driver response model.

Nighttime or Low-Light Studies

When practical for a nighttime pedestrian collision reconstruction, it may be valuable to perform a site visit and a nighttime or low-light study to evaluate site-specific lighting conditions. In doing so, any changes in lighting sources or lighting types between the crash date and the study date need to be noted. As discussed in the "Auxiliary Lighting Sources" section, lighting sources may be changed or replaced with different light types. For crashes that occurred during twilight hours, observations should be made as close as practical to an equivalent time. During civil twilight, ambient lighting will often change rapidly, but the pace of change diminishes during nautical twilight, and especially during astronomical twilight. During the equivalent time, observations on lighting conditions can be recorded. These are often in the form of observations of objects that are visible from a viewpoint of interest that corresponds to a relevant moment prior to the collision. For instance, a position where a driver would have needed to detect the pedestrian in order to avoid the impact. From this position, observations can be made on illumination in the area where the pedestrian

would have been at that time. These observations can include luminance or illuminance measurements, photographs, and/or video.

The use of a contrast reference chart is often helpful in making observations during a study. A variety of reference charts are shown in **Figure 6.9**. Using a reference chart, observations are made during the study of details that are visible under the available lighting conditions. These observations serve as a validation of any visual representation (such as calibrated photographs or video). The chart can be placed at a location where some of, but not all, the details are visible and some at the point of observation, and the analyst can document which details are visible.

Figure 6.9 Landolt C contrast chart (far left and far right), chart based on Ayres 2016 (2nd and 3rd from left), and chart based on Sprague 2019 (4th and 5th from left).

The chart on the far left and far right of **Figure 6.9** was created by the Driver Research Institute. It uses ten Landolt Cs, or broken rings, at various gray shades representing differing contrast levels. Using this chart, observations are made on which of the rings are visible and, if the gap in the ring is visible, which side the gap is on. The second and third charts from the left are based on a study by Ayres and Kubose from 2016 [6.18]. These charts contain eight differing contrast levels, with differing spatial frequencies. Observations are made as to which circles are visible. The fourth and fifth charts from the left are based on a study by Sprague et al. from 2019 [6.19]. This chart contains circles with eight different contrast levels varying in a nonlinear manner to reduce floor and ceiling effects. The order of contrast levels is randomized to minimize observer learning bias, and contrast levels are in an octagonal shape for easy reorientation. Each of the contrast circles also contains a wedge shape, and observations can be made on the directionality of the wedge. These chart types all serve the purpose of aiding observations at the scene and validating visual representations of the scene study.

Calibrated versus Uncalibrated Photographs and Video

Sometimes, a photograph taken just after a crash will be used to illustrate "how dark it was" at the time of the accident or alternatively how "visible" an object or pedestrian was. This is likely to be inherently flawed. While taking the photograph, investigators or witnesses will most likely have their camera (or phone with integrated camera) set to an "auto" mode that will adjust a combination of the aperture, shutter speed, and sensitivity of the camera to minimize the portions of the photograph that are underexposed or overexposed. In other words, the "auto" mode on the camera adjusts settings so that the picture is not overly dark. In doing so, the camera inaccurately depicts the actual lighting conditions. When an accident reconstructionist attempts to accurately portray the lighting conditions taken during a low-light or nighttime study, a procedure is needed to verify that objects of interest that are visible during the study are visible in the photograph or video and, likewise, objects of interest that are not visible during the study are not visible in the photograph or video. Such a process is referred to as photograph calibration or video calibration.

Holohan et al. [6.20] published a technique to "produce a photograph that accurately illustrates the limits of perception for an observer's view of objects under night lighting." The technique involved the viewer observing the signs of varying gray shades during a site study and then producing a range of photographic prints from which the most accurate print could be selected. This process has been updated as photography and display technology has evolved [6.21, 6.22, 6.23, 6.24], but the concept remains the same. The analyst makes observations and/or measurements of relevant objects at a nighttime or low-light study. During this study, photographs and/or video is taken from the same vantage point in a way that emulates the observed lighting conditions while also allowing for adjustments after the study. Then, adjustments are made to the raw media and/or display medium to reasonably match the relevant observations and measurements.

The study by Suway [6.24] discusses a method to calibrate still images and video and display these images with appropriate contrast levels on digital monitors and printed media. This method involves using a luminance meter to take luminance measurements of a contrast board at the scene under conditions with similar lighting. Images or video of the scene and contrast board is then captured and displayed on a digital monitor or printed. Based on the luminance measurements taken at the scene, the contrast of different panels of the board is calculated, and luminance measurements are taken of the same panels on the display media. The image or video is adjusted such that the contrast on the display media is similar to the contrast from the scene measurements. Suway recommends adjusting the contrast of the display media based on whether the object in question is visible or not visible to the observer at the site study.

When displaying calibrated digital media, the use of an organic light-emitting diode (OLED) screen may be useful. OLED screens have an advantage over traditional LED or LCD screens in that they do not require a backlight source to display. Therefore, dark areas of the media can be displayed much darker on an OLED screen than on a backlight screen. This technology works particularly well for displaying accurate calibrated images.

Case Study No. 1: Night Pedestrian Accident with Unknown Impact Location

A collision involving a Ford SUV and a pedestrian was reconstructed. The collision occurred under nighttime conditions with the sun at an altitude of approximately 39.5° below the horizon. The traffic accident report indicated that the Ford SUV was driving southbound and the pedestrian was crossing the roadway from west to east in an area that was not near an intersection. The roadway was dry, and the surrounding area was unlit. The roadway consisted of two northbound lanes, two southbound lanes, and a painted median area separating the northbound and southbound lanes. The Ford was traveling in the leftmost southbound lane prior to the impact. The driver of the Ford stated that there were no streetlights or crosswalks in the area of the impact, his headlights were on, and he saw a black silhouette just prior to colliding with the pedestrian.

Law enforcement officers documented the physical evidence and rest positions with photographs, marked evidence with paint, took measurements, and mapped the evidence and roadway geometry with a total station. The photographs shown in **Figure 6.10** are samples of the photographs taken by law enforcement. These photographs depict the area of the crash, the Ford at rest, contact evidence and damage to the Ford, orange paint on the roadway, and the area of rest of the pedestrian, as indicated by medical supplies on the roadway. Investigating officers documented and painted ABS tire marks with a length of approximately 68 ft leading up to the Ford's rest position. The pedestrian was projected forward and came to rest approximately 39 ft beyond the point of rest of the front of the Ford.

Figure 6.10 Photographs of physical evidence.

As part of the reconstruction, the area of this crash was inspected, photographed, and digitally mapped using a Faro laser scanner and a DJI Mavic 2 Pro unmanned aerial vehicle. GPS-enabled Propeller AeroPoints were placed at the site as ground control for aerial mapping. After this inspection, Pix4D photogrammetry software was used to generate measurement points for the site using aerial photographs and ground control points collected during the inspection. Between the laser scanner and image-based mapping, millions of georeferenced measurements of the geometry of the subject roadway were obtained. The Ford was also inspected, photographed, and digitally mapped using a Faro laser scanner.

To prepare a diagram of physical evidence from the subject crash, a photogrammetric analysis was conducted on several on-scene photographs using camera matching. Mapping data of the subject roadway were imported into computer-modeling software. Computer-modeled cameras were created to view the mapping data from a perspective similar to that shown in the photographs being analyzed. The photographs were then imported into the software and designated as background images for the computer-modeled cameras. Adjustments were then made to the location, focal length, and viewing plane of the computer-modeled cameras until an overlay was achieved between the mapping data and the scene shown in the photographs. When this overlay was achieved, the locations and characteristics of the cameras used to take the photographs had been reconstructed. Once the camera locations and characteristics were known, the physical evidence shown in the photographs was located within the mapping data and placed on a scaled diagram. Through this process, the police paint, the Ford's rest position, and the pedestrian's rest position were located. **Figure 6.11** is the evidence diagram that resulted from this photogrammetric analysis.

Figure 6.11 Evidence diagram.

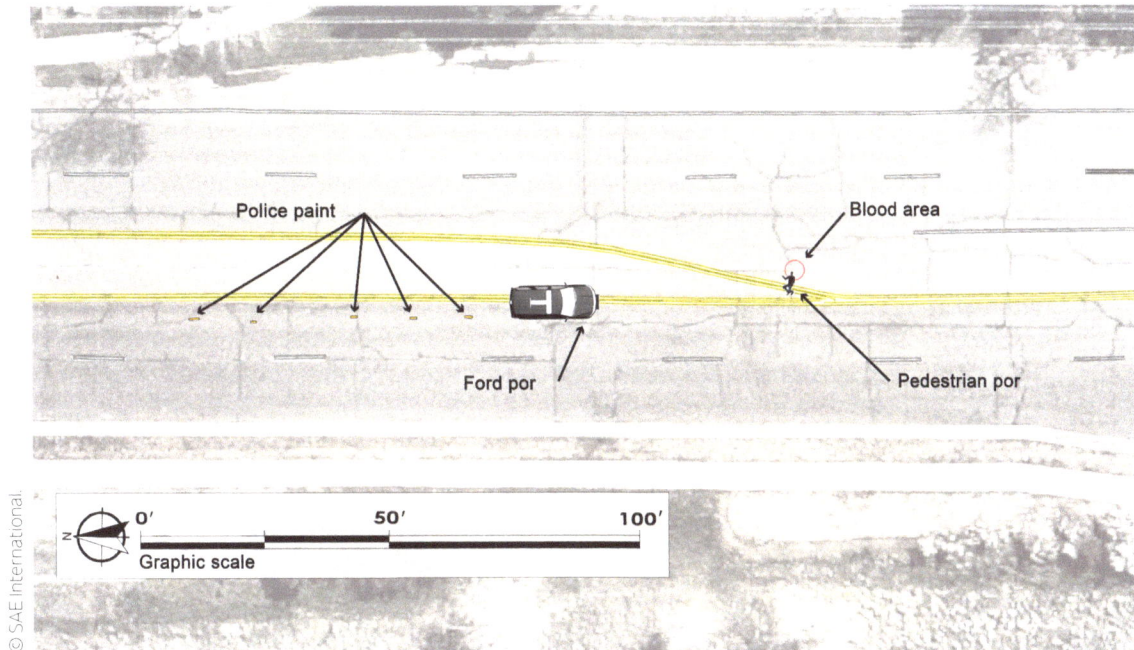

In this instance, the point of impact was unknown. The investigating officer testified that he was unable to locate any shoe scuffs on the roadway. Our examination of the on-scene photographs and video similarly did not reveal any shoe scuffs. Therefore, an unknown impact point analysis was performed. This analysis involved plotting a calculation of the vehicle speed during braking and the vehicle speed based on the pedestrian throw distance and the ultimate separation distance between the pedestrian and the Ford. Ranges were applied to the input variables, and the area where these curves overlapped represented the range of possible values for the vehicle speed at the impact and the location of the impact. Measurements indicated that the pedestrian came to rest approximately 39 ft beyond the front of the Ford. Law enforcement documented 68 ft of ABS tire marks prior to the Ford's rest position. A range of 68 to 78 ft of braking was used to model the Ford's speed. It was undetermined whether the ABS tire marks were deposited by the front or rear tires at the beginning of the mark, so this range accounted for the tire marks beginning with the rear tires (68 ft) or the front tires (78 ft). The braking deceleration of the Ford ranged between 0.76 and 0.80g.

The graph in **Figure 6.12** depicts this analysis. The black lines on this graph represent the speed of the Ford at various positions on the roadway. The position of the front of the Ford from the pedestrian's rest position is plotted on the horizontal axis, and the Ford's speed is plotted on the vertical axis. The horizontal axis intercept of 39 ft represents the distance between the Ford and the pedestrian when they came to rest. The solid black line represents the calculated speed using input values in the middle of the range, while the dotted black lines represent calculations using input values at the edges of the range. In a similar manner, the blue lines on the graph represent a model correlating the vehicle impact speed and pedestrian throw distance. In this case, the forward projection model presented by Toor [6.25] was utilized. The horizontal axis of this plot also represents the distance from the pedestrian's rest position. The Toor model is plotted in blue, with dotted lines representing the 15th and 85th percentile prediction interval. The black lines and the blue lines are overlaid to determine a range of possible impact locations on the roadway and the associated range of vehicle speeds at the impact. This graph includes a highlighted area indicating the overlapping region. The overlapping region indicates that the impact can occur at approximately 80 to 128 ft from the pedestrian's point of rest and that the Ford was traveling between 31 and 43 mph at the impact.

Figure 6.12 Model of Ford's speed at various distances from pedestrian's rest position, vehicle kinematics, and pedestrian throw models with the overlapping region highlighted.

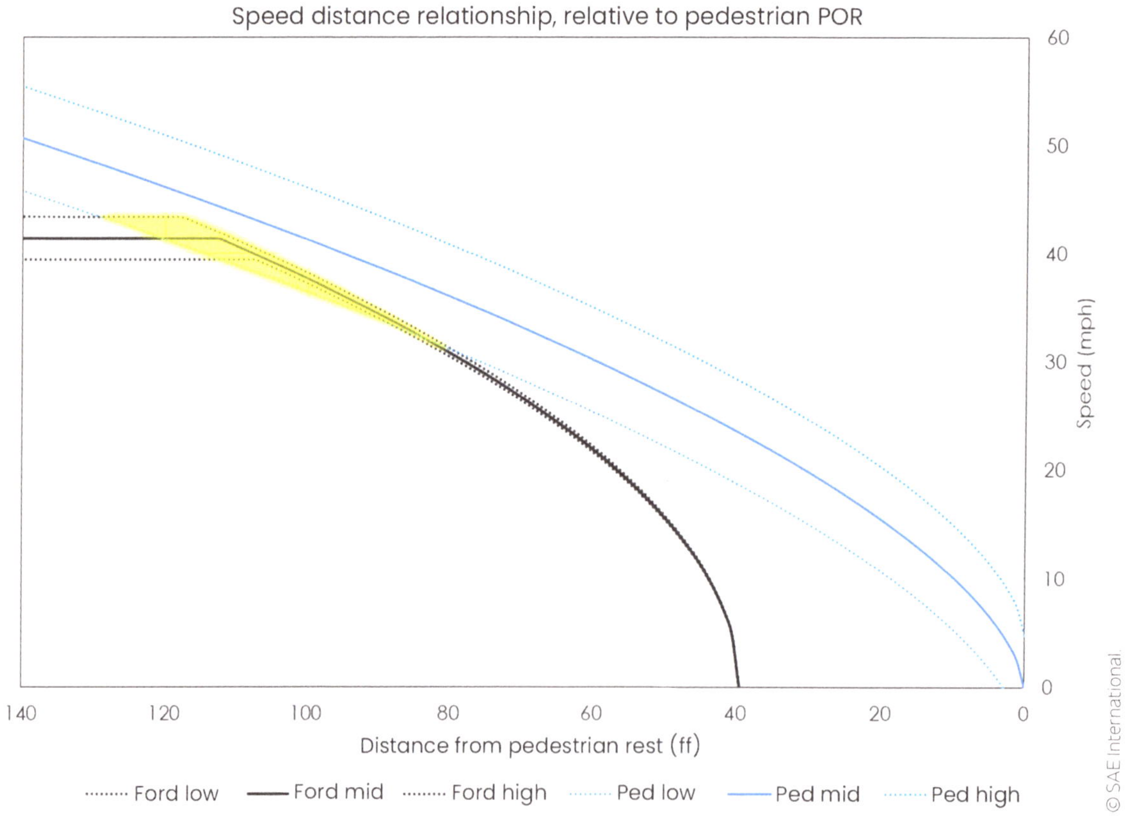

Speed distance relationship, relative to pedestrian POR

To evaluate the distance at which the pedestrian would have been detectable to the driver of the Ford, a nighttime visibility study was performed at the crash site. During this study, the north-bound and southbound lanes were closed, and a surrogate pedestrian stood at the approximate location of the impact. The surrogate pedestrian was of similar height to the subject pedestrian and was dressed in clothing similar to what the pedestrian was wearing at the time of the subject crash. A Sony A7S II camera was mounted on the interior windshield of the subject vehicle. A nighttime scene equivalent contrast gradient panel was utilized at the scene to make contrast observations. The contrast gradient panel uses ten Landolt Cs, or broken rings, at various gray shades. The contrast panel was placed at the scene, and observations were made about which of the rings were visible and, if the gap was visible, which side the gap appeared on.

The subject Ford was positioned at various locations along the roadway, and the visibility of the surrogate was evaluated and documented photographically at each location. At a distance of 50 ft, the shoes and jeans were clearly visible, and the bottom of the jacket was visible. At a distance of 100 ft, the jeans were visible, and the shoes were visible. At a distance of 150 ft, portions of the jeans were barely visible, and at distances of 200 ft and beyond, no portion of the surrogate pedestrian was visible. **Figures 6.13** through **6.16** depict photographs taken at 50, 100, 150, and 200 ft, respectively. These photographs were calibrated to give a fair and reasonable depiction when viewed on a specific monitor with documented settings. However, these images may not provide accurate depictions of the view when viewed in print or on other monitors.

Figure 6.13 Photograph of the surrogate taken at 50 ft from impact location.

Figure 6.14 Photograph of the surrogate taken at 100 ft from impact location.

© SAE International

Figure 6.15 Photograph of the surrogate taken at 150 ft from impact location.

© SAE International

Figure 6.16 Photograph of the surrogate taken at 200 ft from impact location.

After the visibility study, the degree to which the involved parties could have avoided the crash was analyzed. To evaluate the time that it would have taken the driver of the Ford to perceive the pedestrian and begin to respond, the Integrated Driver Response Research (I.DRR) program was used. I.DRR incorporates decades of empirical real-life research regarding drivers and their responses to various hazards. In the time since this analysis was performed, I.DRR has been renamed to Response [6.22]. I.DRR was utilized to determine that the 85th percentile perception-response time for drivers in a situation similar to the driver of the Ford would be 2.5 sec. In those 2.5 sec, the Ford would travel between 145 and 159 ft. Further, it would take between 68 and 78 ft of braking to bring the Ford to a stop. The total stopping distance, therefore, would be between

213 and 237 ft. This indicates that, in order to brake to a stop and avoid the impact, the driver of the Ford would have had to begin detecting and reacting to the pedestrian at a time when the pedestrian would not have been visible. Therefore, the driver of the Ford could not have avoided this collision.

During the visibility study, it was also observed and documented that headlights of the approaching southbound vehicles were visible when the vehicles were well north of the impact location. It was determined that a vehicle traveling at 50 mph would be visible for at least 10 sec. The pedestrian in this case would have had a clear view of the approaching vehicle for at least 10 sec prior to the impact.

Case Study No. 2: Civil Twilight Pedestrian Impact with Glare Source

A traffic collision involving a pedestrian and a Chevrolet pickup was reconstructed. This crash occurred during civil twilight on an unlit rural road, with overcast weather conditions. A driver was traveling northbound when he saw an injured deer in the center of the roadway. He pulled over onto the right shoulder adjacent to the deer and was attempting to remove the deer from the roadway. A southbound vehicle subsequently struck the pedestrian as he was trying to remove the deer. The pedestrian gave a statement that indicated that he was crouched over the deer in an attempt to pull it to the side of the road as the southbound vehicle approached and that he dove out of the way. The southbound vehicle ran over the pedestrian's legs. The pedestrian testified that he was wearing a black jacket with orange non-reflective striping, a black and white hat, and blue jeans. The south-bound driver testified that he saw the headlights from the northbound pickup truck but believed the truck to be in the northbound lane traveling north, rather than stopped on the shoulder. He stated that he heard the thump from the collision after he passed the headlights of the other vehicle, and he did not notice or see anything on the road until after he realized he hit something.

Analysis revealed that the southbound pickup was traveling at approximately 50 mph. In this instance, the headlights of the northbound pickup were a glare source for the southbound driver. Therefore, an evaluation of the view that was presented to the southbound driver prior to the pedestrian contact required an evaluation of the glare produced by the headlights. This evaluation was carried out with a low-light visibility study. A solar-position calculator (suncalc.org) was used to determine that at the time and location of the accident, the sun was at an elevation of approximately 4.5° below the horizon and an azimuth of approximately 248° counterclockwise from due East. Low-light study dates were determined that would allow for similar solar conditions, and a solar equivalent time was determined for each date. Two dates were selected to allow for uncertainty in weather conditions, and the inspection was repeated on each date.

The county officials approved the closure of the road to enact the collision scenario for the study. An exemplar pickup was positioned on the shoulder where the northbound pickup had been parked. The headlights for this exemplar vehicle were set to the auto setting, consistent with testimony. A surrogate pedestrian of similar height and weight to the subject pedestrian was utilized. The surrogate model wore clothing similar to the subject pedestrian, as shown in **Figure 6.17**. During the study, the surrogate was in a kneeling position near the parked pickup, consistent with the testimony.

Figure 6.17 Surrogate model and clothing used in the low-light study.

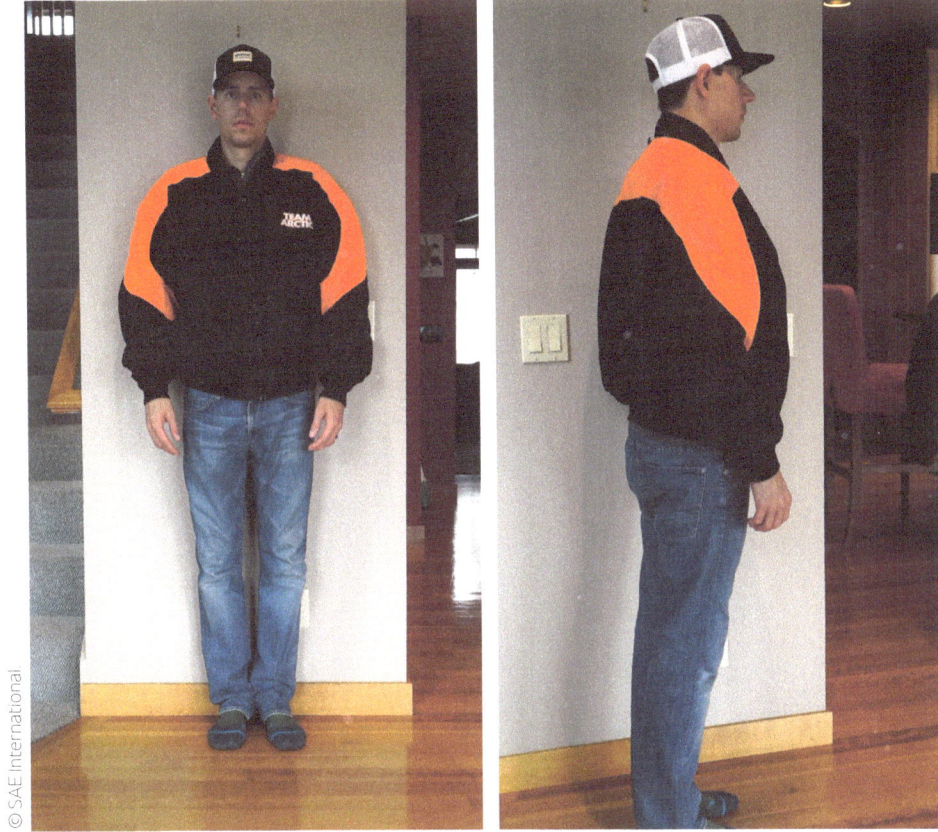

(Continued)

Figure 6.17 (Continued) Surrogate model and clothing used in the low-light study.

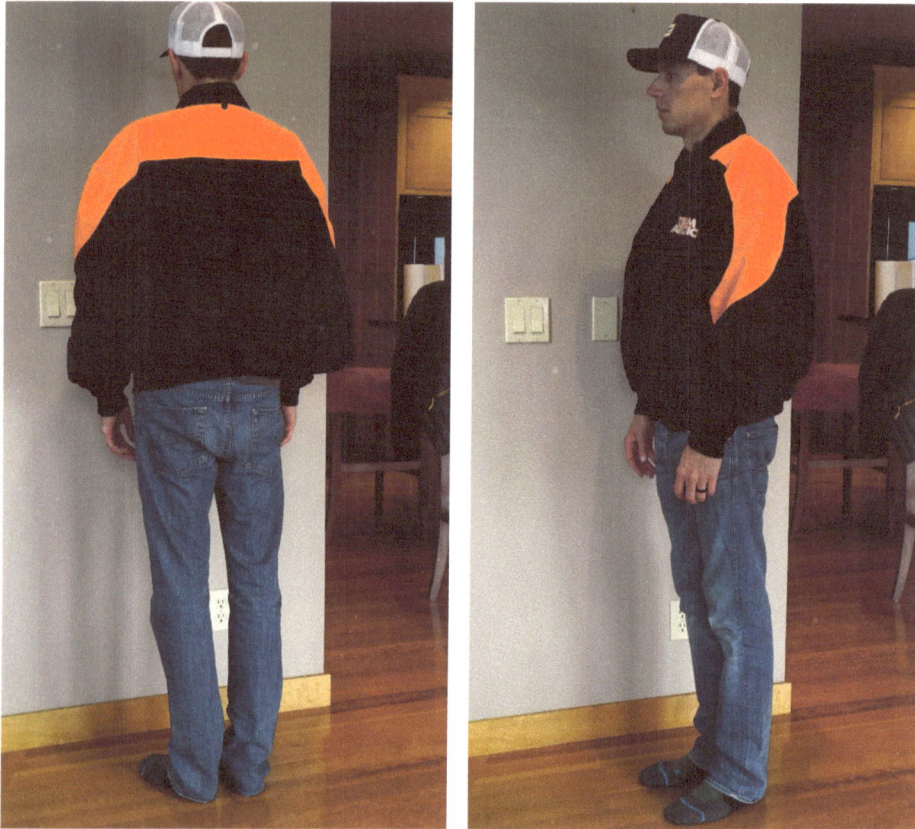

© SAE International

The subject southbound pickup was available for the study, and it was positioned north of the surrogate model in the southbound lane, at a distance that represented the total distance that it would have taken for the driver to detect the pedestrian, apply the brakes, and bring the pickup to a stop. The goal of the study was to evaluate the detectability of the pedestrian at this location on the roadway, with the consideration of the lighting conditions, the glare of the headlights of the parked pickup, and the size, shape and clothing of the pedestrian. For the study, a Sony A7S II camera was mounted on the windshield at the approximate driver eye height. This camera had a 35-mm full-frame Exmor CMOS sensor with a resolution of 12.2 megapixels. The camera body was fitted with a Sony FE 50-mm F2.5 Full-Frame Standard Prime G lens. On the first night of the study, a Westboro Photonics P501U imaging photometer was also mounted adjacent to the camera. This photometer measured luminance levels. The photometer had a 5.0-megapixel resolution and was equipped with a Kowa 25-mm lens. **Figure 6.18** shows the camera and imaging photometer mounted on the interior of the pickup.

Figure 6.18 Camera and imaging photometer mounted on pickup interior for study.

To aid in documenting the study, a nighttime scene equivalent contrast gradient panel was utilized to make contrast observations. The contrast gradient panel used ten Landolt Cs, or broken rings, at various gray shades. The contrast panel was placed at the scene in an area where some, but not all, rings were visible from inside the southbound vehicle, and of the rings that were visible, some but not all gaps in the rings were visible. At the solar equivalent time, with both vehicles and the surrogate model positioned, observations of the contrast panel and environmental conditions were made, and luminance measurements were taken using the imaging photometer. The false color image in **Figure 6.19** depicts the luminance measurements taken during the study. The color scale on the right side of the image indicates the luminance of the color spectrum in candela per square meter.

Figure 6.19 Luminance measurements taken during study.

After the luminance measurements were taken, the exposure settings of the Sony camera were adjusted to take photographs that approximated the visual observations. **Figures 6.20** and **6.21** depict uncalibrated photographs taken during the studies on the first and second nights, respectively. A significant difference between the two studies was the cloud coverage on each night. During the first night, there was sparse cloud coverage, while on the second night, there was full cloud coverage. To qualitatively assess the contribution of the cloud coverage to the illumination of the scene, images shown in **Figures 6.20** (sparse cloud coverage) and **6.21** (full cloud coverage) were taken with the same camera settings, both at the solar equivalent time.

Figure 6.20 Uncalibrated photograph from nighttime study—night 1.

Figure 6.21 Uncalibrated photograph from nighttime study—night 2.

The photograph in **Figure 6.21** was adjusted to achieve an image that was consistent with the observations made during the study. The calibrated photograph, when viewed on a properly adjusted monitor or screen and in a controlled environment, depicts the photograph in a manner that is consistent with the lighting and contrast levels observed during the study. **Figure 6.22** depicts the photograph after the levels were adjusted. Note that the authors are *not* referring to **Figure 6.22** as a calibrated photograph, as it is unlikely that the reader is viewing this on a book or screen that properly displays the image.

Figure 6.22 Adjusted photograph from nighttime study—night 2.

© SAE International

To evaluate the length of the time that the headlights of the southbound pickup would have been visible to the pedestrian, video was taken from the shoulder near the area of impact while a southbound vehicle was driven from a position out of view of the camera. The vehicle was driven southbound at the posted speed limit of 50 mph. The video showed that the headlights of the southbound vehicle were in view for approximately 20 sec from the time the vehicle came around the corner and into view to the time that it passed the approximate area of impact. This video was taken to depict the amount of time that the headlights from an oncoming vehicle would have been visible to a pedestrian standing near the area of the impact. Calibrating this video was not necessary. The lighting evaluation and night study led us to conclude that the southbound driver would not have been able to detect the pedestrian with enough time and distance to avoid the impact. The pedestrian, however, would have had unobstructed visibility of the approaching headlights of the vehicle for approximately 20 sec before impact and could have avoided the accident by exiting the roadway.

References

6.1. Olson, P.L. and Farber, E., *Forensic Aspects of Driver Perception and Response*, 2nd ed. (Boston, MA: Lawyers & Judges Publishing Company, Inc, 2003).

6.2. Adrian, I.W., "Visibility of Targets: Model for Calculation," *Lighting Research and Technology* 21, no. 4 (1989): 181-188.

6.3. Klein, E. and Stephens, G., "Visibility Study – Methodologies and Reconstruction," SAE Technical Paper 921575 (1992), doi:https://doi.org/10.4271/921575.

6.4. Owens, D.A. and Francis, E.L., "Visibility Distance with Headlights: A Functional Approach," SAE Technical Paper 890684 (1989), doi:https://doi.org/10.4271/890684.

6.5. Green, M. et al., *Forensic Vision with Application to Highway Safety*, 3rd ed. (Tucson, AZ: Lawyers & Judges Company, Inc, 2008).

6.6. Muttart, J., Bartlett, W., Kauderer, C., Johnston, G. et al., "Determining When an Object Enters the Headlight Beam Pattern of a Vehicle," SAE Technical Paper 2013-01-0787 (2013), doi:https://doi.org/10.4271/2013-01-0787.

6.7. Driver Research Institute, "Response" Software, Hampton, CT, accessed March 15, 2025, https://driver-researchinstitute.com/software/.

6.8. Funk, C., Vozza, A., and Petroskey, K., "An Optimized Method for Mapping Headlamp Illumination Patterns," SAE Technical Paper 2021-01-0886 (2021), doi:https://doi.org/10.4271/2021-01-0886.

6.9. Funk, C., Petroskey, K., Arndt, S., and Vozza, A., "Vehicle-Specific Headlamp Mapping for Nighttime Visibility," SAE Technical Paper 2021-01-0880 (2021), doi:https://doi.org/10.4271/2021-01-0880.

6.10. Adrian, W.K., "Die Unterschiedsempfindlichkeit des Auges u nd die Moglichkeit ihrer Berechnung," *Lichttechnik* 21 (1968): 2A (in German).

6.11. Adrian, W., "The Integration of Visual Performance Criteria to the Illumination Design Process Public Works Canada," January 1982, 226-256.

6.12. Aulhorn, E., "Über die Beziehung zwischen Lichtsinn und Sehschärfe," *Graefe's Archiv für Ophthalmologie* 167, no. 1 (1964): 4.

6.13. Blackwell, H.R., "Contrast Thresholds of the Human Eye," *J Opt FOC Am* 36 (1946): 624.

6.14. Blackwell, H.R., CIE Report 19.2, Vol. 1, 1981.

6.15. Ising, K., Fricker, T., Lawrence, J., and Siegmund, G., "Threshold Visibility Levels for the Adrian Visibility Model under Nighttime Driving Conditions," SAE Technical Paper 2003-01-0294 (2003), doi:https://doi.org/10.4271/2003-01-0294.

6.16. Roper, V.J. and Howard, E.A., "Seeing with Motor Car Headlamps," in *Thirty-First Annual Convention of the Illuminating Engineering Society*, White Sulfur Springs, VA, September 27–30, 1937.

6.17. Ising, K.W., "Threshold Visibility Levels Required for Nighttime Pedestrian Detection in a Modified Adrian/CIE Visibility Model," *LEUKOS* 5, no. 1 (2008): 63-75, doi:10.1080/15502724.2008.10747629.

6.18. Ayres, T.J. and Kubose, T., "Calibrating a Contrast-Sensitivity Test Chart for Validating Visual Representations," *Proceedings of the Human Factors and Ergonomics Society Annual Meeting* 59, no. 1 (2016): 380-384, doi:https://doi.org/10.1177/1541931215591080.

6.19. Sprague, J., Meza-Arroyo, M., Shibata, P., and Auflick, J., "Enhancing Contrast-Sensitivity Charts for Validating Visual Representations of Low-Illumination Scenes," SAE Technical Paper 2019-01-1009 (2019), doi:https://doi.org/10.4271/2019-01-1009.

6.20. Holohan, R., Billing, A., and Murray, S., "Nighttime Photography – Show It Like It Is," SAE Technical Paper 890730 (1989), doi:https://doi.org/10.4271/890730.

6.21. Ayres, T.J., "Psychophysical Validation of Photographic Representations," in *Proceedings of the ASME 1996 International Mechanical Engineering Congress and Exposition. Safety Engineering and Risk Analysis*, SERA-Vol. 6, Atlanta, GA, 1996, American Society of Mechanical Engineers.

6.22. Krauss, D., "Validation of Digital Image Representations of Low-Illumination Scenes," SAE Technical Paper 2006-01-1288 (2006), doi:https://doi.org/10.4271/2006-01-1288.

6.23. Sprague, J., Shibata, P., and Auflick, J., "Analysis of Nighttime Vehicular Collisions and the Application of Human Factors: An Integrated Approach," SAE Technical Paper 2014-01-0442 (2014), doi:https://doi.org/10.4271/2014-01-0442.

6.24. Suway, J. and Suway, S., "A Method for Digital Video Camera Calibration for Luminance Estimation," SAE Technical Paper 2017-01-1368 (2017), doi:https://doi.org/10.4271/2017-01-1368.

6.25. Toor, A. and Araszewski, M., "Theoretical vs. Empirical Solutions for Vehicle/Pedestrian Collisions," SAE Technical Paper 2003-01-0883 (2003), doi:https://doi.org/10.4271/2003-01-0883.

Index

About the Authors

Nathan Rose

Courtesy of Nathan A. Rose.

Nathan Rose is a Principal Accident Reconstructionist at Explico. Prior to joining Explico in 2022, he was a Principal Accident Reconstructionist at Luminous Forensics, a crash reconstruction, forensic engineering, and forensic visualization firm that he co-founded in 2019. He also held positions as a Principal Accident Reconstructionist and Director (2005–2019) at Kineticorp and as an Engineer (1998–2003) and a Senior Engineer (2003–2005) at Knott Laboratory. He holds a bachelor's degree in engineering from the Colorado School of Mines (1998) and a master's degree in mechanical engineering from the University of Colorado at Denver (2003). Nathan is accredited as a Traffic Accident Reconstructionist by the Accreditation Commission for Traffic Accident Reconstruction (ACTAR), and he has offered expert testimony as a reconstructionist in courts across the United States. He has published numerous technical articles related to vehicular accident reconstruction, and he has published two books through SAE: *Rollover Accident Reconstruction* and *Motorcycle Accident Reconstruction*. You can find him and his blog online at www.nathanarose.com.

Neal Carter

Courtesy of Neal Carter.

Neal Carter is a Principal Engineer at Explico. Prior to joining Explico in 2022, Mr. Carter was a Principal Engineer at Luminous Forensics, a firm that he co-founded in 2019. He also held positions as a Principal Engineer (2018–2019), Senior Engineer (2013–2018), and Engineer (2007–2013) at Kineticorp, another forensic engineering firm, and Seals Design Engineer at ATK Systems (2004–2007), an aerospace company. He holds a bachelor's degree in engineering, with a mechanical specialty, from the Colorado School of Mines (2004) and a master's degree in mechanical engineering from the University of Utah (2007). Neal is a licensed Professional Engineer in Colorado and California and is accredited as a Traffic Accident Reconstructionist by the ACTAR. Neal regularly publishes technical articles related to vehicular accident reconstruction. These articles have been published in the Society of Automotive Engineers *International Journal of Transportation Safety*, in the Society of Automotive Engineers *Technical Papers* series, in *Collision – The International Compendium of Crash Research*, and in *Electric Energy*. Topics covered in these articles include vehicle damage analysis methods, vehicle deceleration analysis, event data recorders, simulation, motorcycle crash causation, video analysis, the use of unmanned aerial vehicles (UAVs) for accident reconstruction, motorcycle dynamics, and driver naturalistic behavior. He is a past organizer for the SAE World Congress Accident Reconstruction session and a peer reviewer for SAE technical papers. Neal has offered expert testimony in court as a crash reconstructionist.

www.ingramcontent.com/pod-product-compliance
Lightning Source LLC
Chambersburg PA
CBHW050906210326
41597CB00002B/44